SOFT-WIRED

How the New Science of Brain
Plasticity Can Change Your Life

Dr. Michael Merzenich, PhD

ISBN: 0989432823
ISBN-13: 973-0989432825

Parnassus Publishing, LLC
San Francisco

Cover Art (Copyright) by Hugh D'Andrade
Jacket Design by Arin Fishkin

CONTENTS

Part Three: Creating "You"

Part Four: The Brain in Retreat

ACKNOWLEDGEMENTS

This book is dedicated to the hundreds of scientists, engineers, medical doctors, therapists, business specialists, and other professionals who have worked on our University of California, Scientific Learning, Posit Science, and Brain Plasticity Institute research teams. I have learned important things from every one of these individuals—as well as from thousands of other scientists whose work has contributed to our understanding of this science over the past decades. Interpretations and conclusions summarized in this book are largely an endowment from these many scientific friends and colleagues.

I especially thank those individuals who most directly assisted in the translation of our research into the forms of help that can better the lives of struggling children and adults out in the real world. William Jenkins, Paula Tallal, Steven Miller, Christoph Schreiner, David Charron, Carl Holstrom, Athanassios Protopapas, Yoram Bonneh, Merav Ahissar, Hubert Dinse, Sheryl Bolton, Bob Bowen, David Blake, Christian Xerri, Srikantan Nagarajan, Xiaoqin Wang, Jeff Zimman, Steven Aldrich, Henry Mahncke, Etienne de Villers-Sidani, Xiaoming Zhou, Thomas Elbert, Nancy Byl, Fabrizio Strata, Sophia Vinogradov, Travis Wade, Tom Van Vleet, Sam Chan, Mor Nahum, Peter Delahunt, Jyoti Mishra, Craig Lewiston, Jeff Dewey, Hyunkyu Lee, Chung-Hay Luk, Wesley Jackson, and Bruce Wexler deserve special mention.

No one deserves greater distinction as a helpmate in these efforts than does my wife Diane. My sisters have called her "Saint Diane" for good reasons. Special thanks also go to my daughters Marghi, Karen, and Betsy for helping me refine and clarify this attempt to deliver one scientist's perspective about complicated issues of brain science to an informed public audience.

Finally, the personal stories of individuals who have transformed themselves or have overcome their special neurological burdens at some point in life by engaging their brains to change for the better are recorded in this narrative. In these stories, names are often withheld to protect individual privacy. Still, all the beneficiaries of this science whose stories are recorded in these pages wanted their personal histories to be told, to encourage those many millions of individuals who—like them—could benefit from brain change, and better understand their potential for personal growth, transformation, or recovery. I thank them for allowing me to tell you about them.

NOTE

The body of science that can be applied to support the conclusions in this book is massive. Rather than provide detailed annotations within this narrative, I have summarized the more readable primary and secondary references that can help introduce you to the several hundred thousand scientific reports—and to other books that are relevant to our subject—at www.soft-wired.com. For the reader who seeks more information or who might wonder about or question the scientific authority of my treatment of this subject, I suggest that you begin to grow your further understanding of this scientific field at that website.

I would also strongly recommend that you visit the websites www.brainhq.com, www.onthebrain.com, and www.positscience.com. Answers to many practical questions that relate to growing, sustaining, or recovering your brain health may be found there.

PART ONE:
YOUR BRAIN IS A WORK IN PROGRESS

1

INTRODUCTION

We are in the early stages of a Brain Plasticity Revolution. That revolution begins with a clearer understanding that the brain's machinery is being continuously rewired and functionally revised, substantially under your control, throughout the course of your natural life. You have a remarkable built-in ability to strengthen and grow the person that you are, at any age. My goal in writing this book is to help you understand how to make the most out of this wonderful natural gift.

Not too many years ago, the average citizen of the world was ignorant about the relationship between physical exercise and health. Anyone who regularly exercised outside of normal work or play activities was viewed as a curiosity, or even a nut. Personal trainers, yoga, macrobiotics, aerobics classes, good and bad cholesterol, Pilates, Tai-Chi, the treadmill, and a thousand other modern testaments to the importance of physical fitness and conditioning were not in our minds, nor in our lexicon. Now, nearly everyone understands that physical health is an important aspect of personal responsibility and overall health and vitality. This book is all about helping you catch the wave of a revolutionary new understanding of the health and wellbeing of the only thing that you possess that is even more important than your physical body: your brain.

Your brain's plasticity is the main source of the refined skills and abilities that account for your capacities at any age in life. Assuming that you are a typical adult citizen of our modern world, brain plasticity-based changes are also contributing to the slow deterioration of your behavioral abilities as you grow older. There are many things that you could and should be doing that could be expected to slow down—and with any luck at all, sharply reverse—those negative changes. Those same activities can also carry you further upward in your capabilities, whatever your current age or status, so you can grow in your achievements, your self-confidence, your happiness, and your life.

In addition to the vicissitudes of normal aging, there are many other opportunities for problems to arise in this most powerful of organs over a (hopefully) very long human lifespan. By the time we reach the eighth decade of life, more than half of us humans have had neurological or psychiatric problems that have required serious medical treatment. Beyond their immediate impacts, these neurological distresses and diseases contribute to our neurological burdens in ways that almost always accelerate and amplify the negative changes that degrade the quality of our lives, and inevitably add their burdens to those that arise from normal aging. It can take a lot of serious work—and perhaps, a substantial change in personal lifestyle—to keep your brain machinery operating at an acceptably high performance level across the many decades of your life.

Of course, because you are reading this book, you presumably belong to the small enlightened class of individuals who already have an inkling that there are important relationships between an individual's neurological health and the ways in which they are engaging their brain across the span of their life. My goal is to help you understand the true nature of the plastic brain, on the path to understanding how the science of neuroplasticity explains both the origin of the person you are, and the better, stronger person that you could be—if you just take this science to heart.

For further explanations and extensive references and citations related to the information in this chapter, please visit www.soft-wired.com/ref/ch01

2

WHAT, ME, WORRY?

An Introduction to Cognitive Decline

"Old age is like everything else. To make a success of it, you have to start young." -Fred Astaire

Have you ever heard the story about the man who walked past the checkpoint on the border every morning, returning home every night on a bicycle? The man aroused the suspicion of the border police, who suspected that he was smuggling something that was hidden in those bicycles—but no matter how much time and trouble they spent inspecting them, they could never determine just what he was smuggling. If you are familiar with the story, you know that he was smuggling bicycles.

Sometimes we search for answers to important questions that have been hiding in plain sight. "Where does the person that I am come from? What could and should I be doing to enrich this life of mine? Why am I slowly 'losing it' as the decades pass by? What could I do that might help me recover from problems that plague my brain and my life?"

People search far and wide for the answers to these great personal questions. Many have a fatalistic view of the origins of their personhood. Some define themselves as a pre-determined, packaged hand-me-down from some unearthly place. Others see themselves as a direct product of their genes, living a life determined by the biological Fates. Still others drift through life tacitly conceding that they shall never understand why they are what they are—or what they could possibly do to alter the direction of their life.

Most of us are in some ways dissatisfied with some aspects of our human performance repertoire. Most of us witness what seems to us to be inexorable functional decline as the decades pass by. We may anxiously search for the magic elixir that just might sustain us, or might help us crawl out of some neurological pitfall. Millions among us come to hold an almost religious conviction that we can eat or exercise or medicate or meditate our way out of the wilderness of our life and older age. Or we simply wait patiently for that stem cell or pharmacological or other heaven- or laboratory-sent miracle to show up, perhaps to save us just before they carry us away to the intensive care facility or cemetery.

In the meantime, the answers to these important questions are increasingly in hand, directly in front of our noses. Contemporary neuroscience has shown us that you come from you. Your brain is plastic. You have the power within, at any age, to be better, more capable, continuously growing a progressively more interesting life. If you're in decline, you have great resources that can help you sustain— indeed, even regrow—your neurological abilities in ways that can help assure that your active brain shall last as long as your physical body. You have powers of re-strengthening, recovery, and re-normalization, even when your brain has suffered large-scale distortions that accompany developmental or psychiatric disorders, and even when it has been physically damaged in any one of the innumerable ways that can befall you in your life.

If you're still alive at the age of 50 and you live in the United States or Europe, the average lifespan extends into the ninth decade of life. Just about every person who is reading this book can optimistically look forward to living past their 85th birthday. You should know, then, that at that age there is roughly a 50% chance that you will be formally identified as senile and demented. Other individuals in that cohort will have memory or other impairments that prevent them from sustaining an independent lifestyle. In that latter case, the medical term is "mild cognitive impairment" (MCI). The only thing mild about it is its name.

You might also be just a little bit discouraged to hear that well before an individual receives an MCI label, their brain is already well down the path of growing the Alzheimer's disease pathology. Alzheimer's signature beta-amyloid crystals and the micro-fibrillary tangles that are killing off brain cells are easily revealed in brain images in about half of normal 70-year-olds. Yet, by their 70th birthdays, only 7% of people have received a formal Alzheimer's or related senile dementia diagnosis. Sadly, the pathological seeds of greater troubles have already been planted in the majority of their 70-year-old brains.

One of my motivations for writing this book comes from watching my own mother Alma, one of the best and kindest of women, travel down this all-too-common end-of-life path. My mom spent four years in the grips of Alzheimer's disease. Like many older individuals, she had been in decline and had lost her independence years before she received the official Alzheimer's diagnosis. How would you assay the value of spending the last ten years of your life vital, happy, sassy, quick of wit, and full of intelligent understanding, as my mother was in her 70 younger, brain-healthy years? I'm pretty certain that my mom would have said "priceless!" Are you taking care of your adult brain in ways that assure that you're getting the most out of your remaining years on the planet?

We all know that there are glorious exceptions to a bleak end-of-life scenario like my mom's. Many of us know talented and exuberant individuals who sustain and even grow their mental abilities and joy of life, who enthusiastically spit in the eye of Father Time. I recently met Avram, a 93-year-old industrial chemist who, in the course of our conversation, told me about a commercial process that he had developed while he was in his 80s. Avram had determined how to profitably produce sugar from waste wood and recycled paper. With a demonstration factory now in operation in Virginia, his trash-to-sugar alchemy may soon make an appearance on your breakfast table. While you might think that this is a surprising achievement for someone of Avram's age, you should know that he really didn't want to talk about the details of that accomplishment at length in our conversation because he was far more excited, now in his 90s, about his new project. Avram has devised a simple, inexpensive way to make ordinary table sugar taste twice as sweet. If he's correct in his scientific studies and production analyses, only a half spoonful of ordinary table sugar (sucrose) shall be required, in the very near future, to make your medicine go down!

What is Avram's secret?! How has this 93-year-old dynamo sustained and even continued improving his cognitive powers? This book is designed to educate you about life strategies that should greatly increase the chances that you (and the other people you care about) ultimately become a nonagenarian paragon like Avram. It is important to understand that living "the good life" is not just about hanging in there by the skin of your teeth cognitively, to a ripe old age. It's also about making the best of every day, week, month, and year of your life— about having a better life at every age. It's never too early nor too late to redirect your life onto a personal path of greater growth and, if necessary, rejuvenation. Our plastic brains provide us with the capability of operating with greater clarity, power, reliability, efficiency, remembering, and understanding tomorrow, as compared to today. Believe me—and take it to heart—that it is in your general best interest to understand how that can be achieved.

For further explanations and extensive references and citations related to the information in this chapter, please visit www.soft-wired.com/ref/cho2

3

THE GIFT OF ADULT BRAIN PLASTICITY

David's Story

David never spoke back. Throughout his childhood he had often seemed to understand what other people said to him, and could usually follow simple verbal instructions. He just didn't talk.

His worried parents did their best to help him. They took him to be probed and scanned and tested and examined innumerable times, in a great struggle to determine exactly why he was so profoundly non-verbal. His parents spent countless hours trying to help him say those first words. Alas, no professional was able to tell them what David's problem actually was—and all the teaching at home could not awaken his voice.

David often appeared to be distracted and disconnected, and his parents and teachers concluded that he was attentionally and cognitively impaired. School wasn't fun for him. Put into Special Education classes, he had difficulty understanding what his teachers said to him—and of course couldn't verbally respond to them even when he did understand. As so often happens to such a child, his classmates cruelly teased him. This, he understood.

In large part because of this bullying, David's parents took him out of school at a young age, and through most of his school years he worked hard at home to educate himself under their patient guidance, using simple workbooks involving limited written and no verbal language.

By his 20th birthday, David had still not spoken a single understandable word. His academic and social development had been a long exercise in personal and parental frustration. While he was a fine young man in other respects, life prospects were unpromising. Still, you must know, it is a wonderful thing to have a loving mom who just won't give up trying to help you, and David had such a mother. Deidre contacted a member of my scientific team and asked if the training programs that my colleagues and I had developed to help language-impaired children might benefit her son. Because she lived nearby, I suggested that she come visit us to determine whether or not David could use—and potentially benefit from—our training approach. In that visit, my clinical colleagues quickly determined that David had a severe auditory processing disorder, and appeared to be capable of using our training software. While David's hearing was normal, his brain only very crudely encoded word sounds. Our training programs had been specifically designed to help a child overcome exactly that class of deficits—although few children in our studies had been as severely impaired as David.

While David willingly engaged in our computer-delivered exercises, no changes in his speech reception accuracy were obvious over the first several weeks of training. Still, with support from his family, David persisted. It soon became more apparent that David's ability to understand what others were saying to him was improving. That motivated him and his parents to keep up with training, working intensely for about one hour every day over a period of many weeks.

At about two months into intensive listening training, something remarkable happened. David began to talk. After another month, he could hold a simple but cogent conversation with his mom and dad. David has been talking to his parents—and everyone else who will listen—ever since.

Now, several years later, in the words of his mom, "You wouldn't know that there was (ever anything) wrong with David. He has residual problems because of all that time he spent in special ed(ucation) classes, but he is doing very well. And having him talk is just a dream come true."

David agrees.

David's life was transformed because his brain—like your brain—is plastic. The machinery of the brain can be revised, to strengthen, or, as in David's case, create new functional abilities, at any age of life. David's speech-understanding and speech-producing brain is physically and functionally different from that receptively impaired and mute brain that he used to carry around in his skull. David progressively rewired his brain in ways that resulted in the recovery of remarkable, fundamental human abilities in his young adult life—the gifts of normal speech understanding and speaking his mind. By his actions (in this case, through computer-controlled brain training), David changed his physical brain. In effect, David has a new, more-powerful brain that can now understand and produce speech. Not many human capabilities are more useful and more valuable in a life than those!

How does David's story apply to the rest of us? Few among us feel we will have to deal with something as severe as David did. Still, however fruitful, happy, and successful (or difficult) your own passage through life, you are also the possessor of great, exploitable powers that enable your achievement of still greater neurological enrichment and personal growth. My first goal is to help you understand the nature of those great powers. If you want to make the most out of your one and only life on Earth, your understanding of brain plasticity science certainly does apply. With that understanding, I'll try to explain to you how you can bring your brain under your control, on the path to a still better, healthier, stronger, safer, happier life.

For further explanations and extensive references and citations related to the information in this chapter, please visit
www.soft-wired.com/ref/ch03

4

MY EARTH IS SPINNING FASTER AND FASTER

Special Challenges for Developing and Sustaining a Modern Brain

Anthropologists have provided us with a broad perspective about the characteristics that distinguish Homo sapiens from other large-brained mammals. One very striking distinction that has been universally recognized is our truly remarkable capacity for behavioral adaptation. All species of mammals are endowed with the ability to learn—to alter their behaviors to fit the demands of their local environments—but our human capacity far exceeds that of all other animals. We can thrive almost anywhere on Earth, we can capture, find, or grow the nutrients required for sustenance, and we can develop and sustain the cultural resources necessary for maintaining vibrant communities. Interestingly, the anthropologists who first described our powerful adaptive abilities identified us as the least "specialized" of mammals. We are not one-trick ponies or always-sideways-walking crabs! On the other hand, our ability to master almost any environment requires remarkable specialization for every individual within the span of each human lifetime. In this respect, we are by far the most powerfully individually specialized mammal. We humans specialize after each one of us arrives on the planet, by revising our own highly plastic brain in incredibly elaborate and powerful ways.

One crucial achievement of our remarkable brain plasticity is the acquisition of that myriad of special skills and the enormous body of special knowledge that is required for our effective operation in that modern sub-culture in the world into which we just happen to have been born. Across the first epoch of our lives, every human child has the basic task of acquiring the skills and abilities and knowledge that provide us (our brains) with a sketchy synopsis of the long history of cultural development of our species. That synopsis stretches all the way back to Homo sapiens' first appearance on Earth about 200,000 years ago. Following that, people moved forward with a heavy dose of tool mastery and information acquisition that relates most strongly to this latest epoch in our native human society. That massive specialization is accomplished in each one of us almost entirely through brain remodeling—by "plastic" changes that revise the detailed operational control abilities of our individual brains and that load them with enormous stores of culturally specific information. By that brain change, our brains are elaborately individually specialized in ways that assure that we can operate successfully in the context of our own very complicated—and by any objective perspective, very bizarre—modern world.

Alas, because humans also collectively and continuously re-shape our environments in innumerable ways, our human worlds have themselves undergone enormous changes over our species' history. We are all well aware of the fact that the rate of change of the tools and social structures in our world is accelerating. In the first years of Homo sapiens, what children had to learn from their forebears was an anthill compared to the Everest that we face in contemporary life.

Nonetheless, those early ancestors were already using one of the greatest neurological advantages of our species: their brain's powerful, natural plasticity. That remarkable plasticity is what allows cultural transmission from society to our individual brains. Unlike other animals, early humans weren't stuck by biological constraints designed to assure their stability and survival as a species. They were able to create many thousands of new skills and abilities, to develop a panoply of tools that could help them far more richly exploit their natural environments, and to develop new, sophisticated mental constructs over the millennia, at any one time in their history conveying the current cultural position effectively to their offspring—and thereby changing, both individually and as a social collective, in a manner unmatched by any other mammalian species, at what is now an incredible rate.

That generation-to-generation transfer of cultural knowledge took a great leap forward each time human societies evolved more complex and specialized skills. From hunting and gathering, Homo sapiens graduated to modern agriculture. From an oral tradition expressed by primitive speech, we elaborated language to produce great works of literature and complex modern media. In a hundred other ways human societies have advanced: from the cave to the skyscraper, from the idol to the cathedral, from the cave painting to the computer tablet, from counting on our fingers to the abstract complexities of advanced mathematics, from sending messages by banging on a drum to sending messages via the Internet, from the one-wheeled cart to the spaceship.

What this means is that like all modern humans, while you're just another Homo sapiens, at the same time you are a culturally evolved creature who richly benefits from an effectively instructed ancestral history. A lot of very complex, highly skilled behavior is not just expected, but demanded of you. Operating with high success in the world—and getting older in it—didn't used to be so darn complicated. Having trouble managing your finances? Math wasn't much of an issue for the older cavewoman. Can't read as fast as you once could? That wasn't relevant for most of human history. Worried about driving a car at night? Not much of a problem in the 19th century. Having trouble downloading podcasts onto your PDA? Not an issue a decade ago. Not tweeting up to speed? An unimagined problem just a few years ago.

The struggles you face as you try to effectively fit into your world are aggravated by three factors that have been growing across our lifetimes. First, modern humans are getting better at extending the span of our lives, decade by decade. Health prognosticators predict that by the middle of this century, the average life will extend well past 100 years of age.

Why does our species live so long to begin with? It turns out that different animals have gene variations that contribute to very significant differences in longevity. Like most social species, our genes are specialized to set up longer-than-average lifetimes. It has long been argued that human children spend so many years "in the nest" because humans have so much to learn (brain plasticity has so much to achieve) in their young lives, and, in parallel, because children—necessarily still under parental control to accomplish this inculcation—take so long to reach the stage of sexual maturity and sustainable independence. Parents' lives must overlap the epoch of greatest behavioral development of their offspring because, it is argued, they serve a crucial role in sustenance, protection, education, and mentorship. Elders contribute importantly to the survival of human clans because they carry the benefits of long experience and provide a resource library (within their brains) for the accumulated skills and knowledge of a stable, enduring cultural tradition. Our long lives were almost certainly selected for, genetically, because elders contribute beneficially to collective Darwinian survival. We have also gotten a lot better at protecting ourselves from a large proportion of the thousands of ways that our lives can be cut short. In ancient societies, not too many individuals lived long enough to achieve elder status. Now, most of us do.

Second, across that long life, we are continuously challenged to modify our skills and world view. We all know that it can be difficult to keep up with the rate of change and the growth of complexity in our workplaces and in the broader culture. This problem was not so great for earlier cultures, where those rates of change in tools, knowledge, and myth were far slower. To cite a simple example, one of our most important tools, the axe head, actually predates our species. Our anthropoid ancestors invented them more than a million years ago. Even after humans arrived on the scene with our very big brains, it took us more than 180,000 years to figure out how to strap a good handle on this most-useful and universal of early tools! Only through another six or seven thousand years of technical development did we sort out how to balance the weight of the axe head by constructing one that was penetrated by the haft. More centuries went by before our ancestors sorted out how to make the first really sharp, honable axe, out of copper, then bronze, then steel. You might remember that all through the course of this research and development period, painfully slow progress led to the production of a really good axe (and its cousins, the arrow, sword, knife, and spear.) These tools were virtually essential to our ancestors for hunting and for war. Culture just wasn't advancing all that fast, in those earlier epochs of our human history.

Speaking personally, I was born in the early 1940s in rural Oregon. My childhood home was without indoor plumbing, we had no modern appliances except a radio, a phonograph, and a hand-wringer for washing clothes. A telephone was an aspiration, and television an unknown. We grew or bartered for almost all of our own food, except flour, sugar, and seasoning. Most of what I find myself doing most of every day in my present life has required the mastery of tools that had not even been conceived of in my childhood. To be continuously effective as a professional scientist and in life, I've had to learn new skills and acquire new knowledge continuously, in a world that is perpetually spinning away, leaving the skills and knowledge that I have mastered earlier in its wake. It's not easy to keep up with the pace of all of these important adjustments!

A third impediment that frustrates our efforts to get the most out of our lives stems from the fact that the brain in most mature or neurologically burdened individuals is itself in decline. Just when we might need our neuro-plastic resources the most, we often discover that we don't remember and don't learn as effectively as we did at younger ages—or as we did before other neurological vicissitudes have arisen in our lives. We are likely less confident, less attentive, more easily distracted, and more readily mentally fatigued and defeated. With a compromised brain, we can expect to have growing problems in producing fast and agile responses in thought, and in action.

It's not easy to keep up, when the brain is slowing down, the world is spinning like a top, and there are still so many years ahead of you in life.

For further explanations and extensive references and citations
related to the information in this chapter, please visit
www.soft-wired.com/ref/ch04

5

THE GREAT AWAKENING

How I Finally Saw the Light

The decision to dedicate my professional life to the study of the brain stemmed from a strong juvenile interest in philosophy. I chose to pursue the study of human nature as my life's work through the investigation of the origins of human behavior—and of consciousness and the "Self"—in the brain. My parents and friends had to endure this immature young scientist calling himself, whenever asked, "an applied philosopher." This was always said with a smile because I knew that it was a thin joke, given the great distance those decades ago between our limited understanding of the brain and our equally specious understanding, from psychology and philosophy, of the core neurological dimensions of our humanity.

Our scientific ignorance in this era can be attributed in large part to a fundamental misunderstanding about the processes in our brains that underlie our personal behavioral development, and account for the really remarkable individual specialization that applies for every one of us. If you think of the million and one things that you know, and that you can do—that no else on the planet knows or does in exactly the same ways—you can begin to understand how unique you are. When I was a young scientist, we brain science nerds had no real explanation for this rather remarkable capacity of Homo sapiens. Most scientists believed that the brain reached a mature physical and functional state largely in the first year of life. With the completion of early maturational processes, the wiring of the brain was established, and all of the nerve cells (neurons) in the brain achieved their mature ("adult") forms. From the end of that very early "critical period" of development looking forward to the end of life, brain wiring and the neurological elements of the brain were thought to be immutable. Imagine your brain as a set of electronic components and loose wires that, over the first year or two of life, hardens into something like the computer sitting on your desk, and you get the picture.

You might note that according to that point of view, a person's brain really had only one way to go—from its physical status at the end of the critical period in the baby or toddler on to the end of the person's natural life: downhill. And by that view, an individual was stuck with that brain that grew to "physical maturity" pretty much before they were toilet trained! Our abilities, our intelligence, our attitudes, our prospects for a useful or wasted life, our great or minuscule chances for significant achievements were pretty much set in stone well before we ever showed up at the schoolhouse door!

Nothing could be further from the truth.

This egregiously flawed scientific concept arose in large part as a conclusion drawn from studies conducted in brain areas receiving information from the eyes. Those areas were easily altered by manipulating information coming from the eyes to the brain in the first weeks or months of life—but the neurological patterns of representation of information from your left and right eye were hardened to a "mature" form by early visual experiences. Studies indicated that all it took to functionally and physically stabilize the neurology of this important brain area was to just keep one or both eyes open for a while. The basic distribution of activities related to seeing seemed to be easily changed or distorted in the earliest days of life. For example, by closing one eye while leaving the second eye open through this infant epoch, the brain machinery "representing" the open eye expanded, while the machinery representing the closed eye contracted. However, once the brain had "matured" across this "critical period," no manipulation of vision from the eyes could induce further changes.

We now know that information combined from the two eyes must be cast in concrete in early life because very precise two-eye combinations control a critical facet of our seeing—the brain's reconstruction of our visible world in three dimensions. Seeing in 3-D requires that the brain accurately reconstruct the differences between what you see with your left eye versus what you see with your right. The brain's processes that account for setting up binocular (two-eye) vision are necessarily just about the least "plastic" of the change-control processes in the brain.

Most unfortunately, the study of the hardening of this system, conducted by brilliant research scientists who later won a Nobel Prize for this and related work, provided brain scientists with their most complete and most influential model for documenting postnatal brain development. When scientists in this field "discovered" that the brain was immutable after only a few weeks or months of life, the conclusion that the brain was inalterable at older ages and throughout the remainder of life was quickly adopted as a generally established principle for mammalian brain organization.

How could a hard-wired brain change its performance-control operations in ways that could explain the incredible evolution in human abilities expressed all across our childhood and our juvenile and adult lives? Explanations were necessarily hand-waving. Our computer-like brain obviously had to have a pretty wonderful—and wonderfully mysterious—"operating system!" (Only) God knew how the devil it could actually work! Still, despite its scientific implausibility, most scientists, medical doctors, and professional therapists believed that brain functions were strictly and permanently localized in a brain that was hard-wired from a very young age.

In my early research career, I had focused on hearing, language, and the processing of information coming from the physical body. I began to realize that what I was witnessing in hearing, language, and motor machinery in the brain did not fit the predominant perspective coming from the hard-wired visual system model of how the brain works. One challenge arose when we began to document individual differences in how sounds were represented in the brain, or in how sensations from the skin of an animal excited the brain in different individual animals. One adult monkey might have a highly detailed pattern of representation of the thumb on its dominant hand that was many times larger than the very crude representation of the dominant-hand thumb in a second adult monkey. One adult animal might have very elaborate and refined machinery representing environmental sounds. A second animal might have greatly simplified and only much cruder machinery representing the same repertoire of sounds. "How do these differences in wiring arise?" I asked myself. Surely such differences must reflect an individual animal's learning experiences. Surely they must reflect an individual animal's special abilities and competencies. Surely those competencies can grow or wane, as we know our human abilities can wax or wane across our lifetimes.

The breakthrough came from a study conducted more than 20 years ago, when my colleagues and I documented how the adult brain recovered from a sensory hand injury. Let's imagine a large area of your palm was cut or severely damaged. The prevailing view at the time suggested that the region in the brain representing the palm's surface would henceforth remain silent and unused. It was widely believed that the adult brain should not be able to revise its wiring (or "maps") of the skin surface of the hand after such an injury. What our breakthrough study found was, in fact, the exact opposite; following such an injury, your brain rapidly and dramatically revises the way that it represents your skin surfaces. After injury, intact skin surfaces represented in the territory surrounding the zone of loss rapidly moves in, to occupy the "abandoned" region. After a long period of neurological remodeling—which occurs hour by hour, day by day—inputs from the surrounding skin surfaces come to be represented in enhanced detail in the brain zone formerly representing the now missing skin.

What happens when the cut is healed and the nerve regenerates? Again, the brain areas that represent the skin undergo a process of remodeling and change. Throughout the weeks it takes to heal, the regenerated nerve inputs try to reclaim their old neurological territory, which has been taken over by neighboring parts of skin. Over time, the healed nerve will reclaim much (but never all) of its previous brain territory, which will now represent the reconnected skin in a dramatically new way.

The expressions of change that we witnessed in these territorial "brain wars" did much more than simply demonstrate the fact of adult brain plasticity. They also revealed the fundamental nature of the neurological processes that control brain remodeling. Most of all, they showed that the representation of the surfaces of the body in your brain and the wiring that sustained them are not cast in concrete. To the contrary, they were continually revisable.

Not *hard-wired*, but *soft-wired*.

It did not take long to extend these studies into the realm of behavior. Our strategy was to train animals in ways that led to their development of new skills or abilities while we tracked brain changes that might account for that skill acquisition. It turns out that these are among the easiest of studies to conduct in all of brain science. If we scientists looked in the right way in the right places, neurologically speaking, skill acquisition or performance improvement at any skill or ability was brain change!

Of course scientific discoveries like these breakthrough studies did not occur in a vacuum. There were many important antecedents to the work, and many other scientists throughout the world contributed to the development of our understanding of plasticity processes. Historically, most scientists in the late 19th and early 20th centuries had actually accepted the modifiable ever-changing-brain as a given. Even while the predominant view shifted in the middle of the century to the claim that older brains were incapable of change, there was a small group of brain scientists, the "physiological psychologists," who were busy documenting many aspects of brain plasticity. Sadly, almost no one in the mainstream camp (myself included) paid enough attention to them. One limitation of their studies was the almost exclusive use of a Pavlovian conditioning model. By that behaviorist model, the involuntary association of a stimulus with a reward or punishment resulted, as the great Russian scientist Ivan Pavlov showed us early in the 20th century, in the neurological establishment of that stimulus-to-reward (or punishment) relationship, evidenced by a behavioral response that anticipates the reward (or punishment). Pavlov's dog heard a bell, then, a little later, received a small chunk of meat. After one or a few trials, every dog had learned that the ringing bell predicted the meat. Pavlov's dogs showed that they had established that neurological association by their obvious impatience and anticipation for the meat's arrival every time that bell rang.

Scientists that you may never have heard of, like Charles Woody, Jerome Engel, Richard Thompson, E. Roy Johns, John Disterhoft, Norman Weinberger, and a host of Russian scientists (among others) had conducted compelling studies documenting large-scale changes in brain responses that met all of the contingent requirements of Pavlov's model. For example, the UCLA-based Dr. Woody and his research fellow Jerome Engel recorded large-scale changes in the neurological representations of both the brain's recording of a punishment (a blast of air) and a conditioned response (a twitch of the nose), in rabbits conditioned to avoid aggravating puffs of air delivered to the nose. John Disterhoft first showed that Pavlovian conditioning-induced changes amplified and increased the selectivity for the representations of specific "conditioned" bell-like sound stimuli. Norman Weinberger's group at the University of California at Irvine repeated and elaborated those studies, recording single-brain-cell changes in responding for specific bell-like stimuli that endured as long as the behavioral association between that sound and a punishment or reward was sustained—then reverted to a pre-training status when "behavioral conditioning" was "extinguished"—in the latter case by repeatedly delivering a sound stimulus with no associated reward or punishment. Dr. Richard

Thompson and colleagues (then at University of California at San Diego) showed that all of the changes contributing to Pavlovian conditioning could be documented to occur within a given local brain center. In parallel with these studies, scientists like Mark Rosenzweig and Marian Diamond at the University of California at Berkeley and E. Roy Johns at New York University had shown that plastic remodeling is a near-universal property of the mammalian forebrain. Train an animal any which way following the "rules" of classical (Pavlovian) conditioning or radically "enrich" an animal's living environment, and watch its brain revise and elaborate its neuronal connections. In the light of these important earlier studies, I have sometimes been very embarrassed to be called "the father of cortical plasticity."[1]

At the same time, my research group did extend these earlier studies in a number of important ways. We helped document the facts and forms of plasticity in the domain of operant conditioning—the progressive improvement of voluntary responding as we acquire or refine our skills and abilities—which is a primary route of human skills acquisition and improvement. We illustrated the nature and power of brain remodeling in studies of nerve and brain injury that bore seminal implications for rehabilitative medicine. We provided additional proofs that the brain organized itself on the basis of the concurrent arrival of information in small moments of time, which helped bring plasticity research onto a higher logical plane. We described "adult" plastic changes in terms that helped scientists better understand how they actually changed physical and functional structures in the brain. We added substantially to the developing understanding of how plasticity was controlled and regulated in adult brains. We provided compelling evidence that helped us understand when and how our selective attention, our memory, and our powers of prediction controlled adult cortical plasticity. We contributed to studies that demonstrated that plasticity could not be isolated to any single brain system level—but to the contrary, as we acquire new skills, remodeled entire brain systems and networks. We added to a growing body of studies showing how plasticity and its consequences change as we progress from infancy to childhood to adolescence to young-adulthood to mature-adulthood to end-stage epochs of life.

---

[1] The word *cortical* refers to the cerebral cortex, the thin sheet of brain cells that covers the hemispheres of our forebrains, to which we attribute many of our human powers of perception, cognition, memory, and the control of our actions—and where scientists generally locate the "person" at the helm of the ship!

21

Perhaps most importantly, we began to demonstrate how these processes might be employed therapeutically to empower under-functioning brains, or to drive neurologically or psychiatrically impaired or dysfunctional brains in strongly corrective directions through training.

When my own brain woke up to the fact of adult plasticity and to the many implications it held for strengthening and recovering brains that could benefit from or were in need of a boost or help, I thought that the scientific and medical communities would quickly correct their wrong-headedness and rapidly exploit this new understanding. It turned out that there was a long road ahead, for remodeling the brains of all those brain scientists who were stuck with a bankrupt notion of how their brain machinery accounted for their own behaviors. This revolutionary scientific perspective was initially received with skepticism, and subjected to substantial ridicule. Of course, in the end, correct ideas win out, and now, most scientists understand that the brain is continuously plastic throughout our natural lives.

For further explanations and extensive references and citations related to the information in this chapter, please visit
www.soft-wired.com/ref/ch05

6

THE INCREDIBLE LIFELONG POWER
OF BRAIN PLASTICITY

Lessons Learned from the Development
of the Cochlear Implant

As a young scientist at the University of California at San Francisco (UCSF), in parallel with the just-described studies on the functional organization of the brain that led to my scientific awakening about brain plasticity, I led a team that created one of the world's first cochlear implants to help deaf people regain hearing. The cochlear implant is a surgically implanted device that electrically excites the auditory nerve via a long thread-like array of stimulating electrodes. Our cochlear implant was designed to generate patterned electrical activity across the surviving hearing-nerve wires in a profoundly deaf ear in an attempt to mimic the normal patterns of activity that would represent normal human speech. The better that simulation, we told ourselves, the more likely it would be that a deaf individual would interpret that electrically patterned activity as normal speech.

The production of a practical clinical device was not that easy to achieve. Implanted electronics must last a lifetime. The implant had to be safely introduced into the fragile hearing organ. The hearing nerve and the appliance had to survive a heavy schedule of electrical activation over a period of many decades. These implanted instruments turned out to be the most sophisticated electrical devices implanted in the human body up to that time. Moreover, electrical stimulation in this human tissue was difficult to control, and the patterns of activity delivered from the ear to the brain with our devices only crudely simulated the elegant, refined patterns produced by an intact inner ear. Shocking the spiral-form hearing nerve with four or eight pairs of wires in our early devices was akin to playing Chopin with your forearms or fists!

Not too surprisingly, when patients first heard someone talking as they listened during the early days of their cochlear implant use, most found the voices that they heard to be unintelligible. "It sounds like a robot," a patient might say. "It sounds like a radio that's not tuned in to the station," said another. "It sounds like crap," said a third.

We were not surprised by this reaction. After all, the electrically coded signals that were being fed from the ear to the brain were very different from the far more refined patterns of hearing nerve activity that the patient received before they had lost their hearing. It was difficult to span the hearing nerve sector that is the primary source of aural speech information in the normal way. Areas of the nerve stimulated by different electrode elements complexly overlapped and interfered with one another in the ear and in the brain. The elegant distributions of activity representing sound in fine detail in time, believed by auditory neuroscientists to be crucial for encoding the tonal, timbre, and loudness qualities of sounds in the normal ear, were grossly degraded as we attempted to represent them by so crudely electrically shocking the nerve.

In one of the first patients that we studied, we organized a special experiment to try to understand how we could improve our speech-coding strategy. Earl was a man of about 60 who had lost his hearing over the preceding several years. That loss had been devastating to this formerly very social individual. While he was deeply discouraged by his inability to communicate with his friends and family members, Earl reliably showed up for testing in good humor, with cartoons that he had drawn usually depicting we scientists and doctors in comic situations, or revealing to us what Earl thought about our foibles—which, in our family of nerds, were rich and varied—in hilarious terms. We struggled to communicate with this wonderful fellow, we were terribly fond of him, and we were all motivated to help him.

One bright idea was to ask Earl to report exactly what he heard as we delivered lists of words to him via his cochlear implant. We had reasoned that an analysis of what he said he heard, sound by sound, might guide us to do a better job of speech coding in the electronics of his implant. As we delivered each word, Earl thought for a minute, then uttered his answer. For example, if I said "car," Earl might respond (with his voice) that he heard "kera." If I said "box," he might respond "brruh." If I said "damp," he might say "aahema." While we pored over his responses at length, they were very discouraging, because it was almost impossible to reconstruct consistent, correctable patterns of distortion in the cochlear implant-coded speech that Earl described.

However, not too long into this study, I had a second bright idea. I opened the heavy door of the sound-proof experimental chamber where Earl was sitting for this testing, and gave him new instructions. Handing him my word list, I said, "Earl, let's try something different. I want you to read aloud each word on this list. As you read it, listen carefully. Then tell us exactly, in your own voice, what you just heard yourself say." Over the speakers in the control room I heard Earl say "car;" then, after a significant pause, we heard him say, slowly, "caarr." Then Earl said "box." Followed by a slow "box." Excitement was growing in Earl's voice as he said "damp," now rapidly followed by "damp." By the fifth or sixth word, he couldn't contain himself, and insisted that we come back into the test room so that he could "tell us about it." With great excitement, Earl enthusiastically told us that "all you have to do is to make everyone sound just like me talking, and I'll understand everything!" Alas, we knew that the acoustic differences in Earl's voice and other voices could not possibly explain why Earl heard such distorted and un-interpretable sounds when someone else was talking with him, but "heard everything in a completely normal voice" when he did the talking. In our initial studies, we worried that we would be able to provide only limited hearing benefits for a struggling individual like Earl—and that he would have to undergo potentially many years of training to be able to understand the very strange foreign language that our cochlear implant was providing for him.

Just imagine our great delight—and the great delight of the developers of other commercial cochlear implants that would evolve from scientific studies conducted in other laboratories during the same era—to have our patients tell us some months later that they now heard and understood almost everything being said to them! With the commercial devices produced by Advanced Bionics that ultimately brought our UCSF laboratory's science out into the clinical world, that wonderful outcome was reported in more than 80% of formerly totally deaf individuals. After these months of device use, patients commonly reported that when they now heard someone else talking that "it sounds just like it did before I lost my hearing. Their voice now sounds perfectly natural."

There is one other really fascinating aspect of this wonderful outcome that knocked me for a scientific loop. The other cochlear implant teams in other research centers of the world initially employed different strategies for coding human speech in their devices. They also translated those very different codes to electrically shock the hearing nerve in fundamentally different ways. To our astonishment, it did not make much difference. How could all of these very different versions of cochlear implants work so well? The brain just didn't care! The majority of patients implanted with these very differently operating devices also recovered speech understanding—and many of their patients also came to describe the speech that they heard with their cochlear implant as sounding "completely natural."

It took me awhile to realize that the success achieved with these devices was not directly attributable to our clever engineering. All we scientists and engineers had done, in a sense, was to provide information about the complex sounds of human speech to the brain in several different, crudely patterned ways. The plastic brain did most of the real work. It made sense of our crudely encoded sounds. It re-organized itself, to create a new, significantly different neurological reconstruction of translated sounds that our patients identified as normal, understandable speech. Moreover, these devices could all work beautifully to establish speech and language abilities in an eight-month-old, or to restore speech understanding and language usage in an 80-year-old. Although it was completely unplanned, studies with the cochlear implant turned out to be the grandest demonstration of the power of lifelong brain plasticity conducted in all of brain science up to this present time. We were witnessing fabulous plastically adaptive brain power!

I learned another important lesson about brain plasticity from the study of patients like Earl. Earl had lost his hearing in his 50s, over a period of several years. He was an educated fellow with a sophisticated history, and as with almost all individuals who acquire deafness at an older age, had retained the ability to speak, albeit in a distorted manner. There was no question that he had all of the neurological gifts of "internal language" prior to receiving his cochlear implant. As I have described, when he spoke, his brain told him that what he said and what he heard himself say were the same. Speech understanding is rapidly reacquired in individuals like Earl, I believe, because their own voices are their teachers. His remembered speech production abilities powerfully shaped the remodeling of his speech reception machinery by informing his brain a million or so times a year that the words he heard when he spoke were correct. Through that teaching, his brain ultimately re-established the correspondences between its representations of his self-confirmed words and the same words produced by the voices of others.

When patients like Earl later understood the speech of everyone else, I quizzed them on how they employed their new speech understanding to access all of that information that they had recorded in the brain from that long epoch of their younger lives when their hearing was still intact. I'd usually engage them in a conversation about things that I thought would be special in their individual pre-hearing-loss experiences. Because Earl was college-educated and lived in a rice-growing region in California's Central Valley, for example, I might have chosen to talk with him about his college major, or about rice farming and the rice business.

Whenever I had such a conversation, I saw that the patient had reconstructed a seamless reconnection between their completely reorganized speech-representing brain machinery and all of those great stores of knowledge—their memories—that they had recorded earlier in life, when their recording machinery was fed by their old, very different, pre-deafness brain machinery. Now ask yourself: "If I radically change the detailed sources and forms of information feeding the brain's language system so that it has to make a trillion or two changes in wiring to again represent speech as intelligible, how can it make all of the correct adjustments to make all of those tens of thousands of words and all of those millions of word associations and all of those many millions of retrievable moments of language-based memory be so exactly, correctly reconnected?" Those adjustments would presumably require that the brain make hundreds of billions or possibly trillions of new connections! The recovery of full access to a massive body of earlier pre-deafness speech-referenced knowledge through cochlear implant use blew a hole through the adult-brains-are-hard-wired claim that you could drive a Mack truck through.

The obvious conclusion from these observations was that memories must control the re-establishment of this new speech-reception machinery. This demonstration that plasticity in a brain system can be magnificently guided by memories, from the top of one of our great brain systems, has been supported by a growing number of studies that show that remembered neurological constructions of the things of the world actually control and shape our brain plasticity. In chapter 31, I will explain how this important principle has influenced our designs of brain exercises and activities that can be expected to most efficiently and effectively improve the operations of your brain.

It has been a great privilege for a scientist like me to work on the development of a device that has transformed the lives of many, many thousands of people for the better. Restoring useful hearing in an individual who has acquired profound deafness is, after all, the stuff of miracles.

At the same time, my team's work on the development of these devices gave us powerful new insights into how the listening and speaking brain must really be operating. It greatly reinforced my growing understanding, as a young scientist, that the adult human brain is powerfully self-adjusting. It's soft-wired, not hard-wired. The cochlear implant is not a miracle of neuroscience-guided engineering. It is, rather serendipitously, a manifestation of a truly remarkable adult capacity for brain plasticity. As my colleagues and I (and the members of other cochlear implant development teams across the world) declared victory over acquired deafness more than 25 years ago, even then I understood the greater implications of the recovery of hearing in these patients:

*Human brains are **fundamentally** plastic.*

For further explanations and extensive references and citations related to the information in this chapter, please visit
www.soft-wired.com/ref/ch06

7

TRANSFORMING LIVES

It's All About Your Brain and Your Future

They paddled John at St. Michael's School in Santa Fe because he couldn't read. Every child got their turn striking the back of his legs before John was sent ignominiously back to the seat that he occupied in the "Dumb Row." Back there, he could continue to stare at the meaningless squiggles that seemed to make sense to every other kid in his class—but not, alas, to him.

No one wanted to read more than John did. Like most children who struggle with learning disabilities, he was ashamed of his failure and did absolutely everything possible in his power to disguise his problem. He was extremely adept at behaving in a way to avoid drawing any attention to himself that might reveal his disability, and so good at hiding his abject failure and the cheats he created to compensate for it that even his parents and closest friends did not realize how utterly and totally he struggled as a reader. Truth be told, John couldn't read a billboard, street signs, headlines—in fact, John couldn't read anything. Miraculously, John survived in school because of an exceptional ability to remember what he heard, and by aggressively using friends and other students as helpmates in "discussing" his reading assignments, in "helping him" write or type required written reports, by using every trick under the sun to escape written exams, and by becoming a master at cribbing off of other kids in class. It helped that John was intelligent, a superb athlete, personable, positive, and easy to forgive. Still, John was like a duck on the water—calm on the surface, but paddling furiously out of view. John lived in continual fear that his friends or teachers or family would discover the profound level of his impairment—which he believed would mark the day that his happy life would come to an end.

How far can you advance in an American school system when you can't read at all? John was helped in advancement because he was an athlete, and that undoubtedly eased his progression through junior high and high school, his graduation from a junior college, then perhaps astonishingly, from the University of Texas at El Paso. With absolutely no ability to read, what could a smart young college graduate like John possibly adopt as his life's profession?

John became a high school teacher.

John got the job in part because he had his father fill out his application, which he verbally dictated to his dad over the phone. When John met with his new boss several weeks later, he discovered that one of his first assignments would be to teach English grammar! Just at a time in life when he may have walked remarkably unscathed out of the minefield, John voluntarily re-entered it, back where the risk of discovery—perhaps the greatest threat to his mental and emotional life and limb—was even greater!

Most facile readers do not know how heavy the burden of shame can be for the non-reader. I once had a conversation with a gentleman who then ranked among the 100 wealthiest Americans, when the subject of reading failure in school-aged children came up in conversation. A tear glistened in his eye when he was reminded of his own childhood dyslexia. Children who struggle with reading may think, "Why can't I do something that most other kids can do? What's wrong with me?" These questions are difficult for any child to escape, and many find themselves repeating them a thousand or ten thousand times. Those questions, and the shame behind them, reside in the recesses of the mind of most non-readers for the rest of their lives.

John has beautifully described the personal burdens he carried into his classrooms as being "the schoolteacher who could not read."[2] He operated in continual fear that his students and professional colleagues would discover his learning deficit. No time was spent at the blackboard in John's classes! He became an expert at having his students read their assignments and educational materials to one another—in effect, teaching themselves. He spent lots of time listening to teaching materials on tape, and in talking with his fellow teachers about their lesson plans. Perhaps the fact that no essay assignments were made in his classes added to his popularity as a teacher!

John spent more than a decade as a tenured teacher at different

---

[2] *The Teacher Who Couldn't Read* by John Corcoran.

public high schools in Southern California. He was well-regarded by his superiors as an engaging and effective teacher and coach. His wife was the only person in his world who knew that he could not read a word. It was only after a successful departure from education into a business career, during a period when his business was in the doldrums, that John finally decided, once and for all, to try to kill the demon that he had so long been carrying around with him. Near his 48th birthday, he entered an adult reading recovery program in San Diego. Starting back at the beginning, as if he were a baby again, under expert professional tutelage, John gradually re-acquired all of the language and visual skills that support reading, then slowly mastered reading itself. Not too many months into that program, John could read. He could read the newspaper. He could read the telephone book. He could read magazines and novels. He could read everything.

John now reads voraciously. He reads for fun. He loves to write. John has been transformed, through adult brain plasticity. John is no longer ashamed.

You would be a rare individual if there were not things in your life that have held you back—sometimes, more than a little. John's experiences show us, once again, that transformation in your neurological abilities is not just the stuff of childhood. Even the most fundamental of problems can potentially be addressed at any age. Like John, you (and all those people that you care about) can improve at almost any ability with appropriate forms of brain training. After all, your brain is just as plastic as John's brain!

Józef Korzeniowski was an orphan who, as a lad of 16, left his home city to travel to Marseilles to begin the life of his childhood dreams. He signed on as a seaman on the crew of a small Mediterranean sailing ship. Drinking, carousing, smuggling, and fighting—by all accounts, Józef enjoyed his life at sea and at every port of call. Still, his sense of exuberance and fun were often tempered by epochs of self-doubt, cynicism, and depression. By age 20, with some help from an uncle, Józef had built up a small nest egg. When his life savings was stolen from his room, Józef, like so many individuals who oscillate between the highs and lows in life when they feel largely alone in the world, turned a pistol toward his chest and pulled the trigger. Fortunately, young Józef was unschooled in the science of human anatomy, so the bullet passed through him without hitting the heart or other vital organs.

What great good fortune each one of us is entitled to, at least a few times across the span of our lives! Gradually recovering his pluck, Józef returned to sea with even greater determination to make a go of life, rapidly gaining his credentials as a second mate—and over a several-year period, advancing to the rank of Captain on a small merchant sailing vessel.

Near the end of the 19th century, although commercial sailing was fading as an occupation, Józef was clearly in his element on the deck of a ship. He had a wonderful memory for the strong and interesting characters that he met along the way, in the great oceans, seas, bays, and rivers of the world. It was perhaps a minor tragedy, then, that his health began to fail him in his mid-30s and it became clear to him that he could no longer sustain the physical rigors required to continue at the helm of his ship. On the other hand, Józef had always held the rather audacious belief that he might have the ability to succeed as a writer. Once land-bound he turned his energies toward writing a tale set on a familiar passage along the East coast of Borneo. Perhaps more than a little surprisingly, this first novel, Almayer's Folly, was rapidly published, and actually turned out to be a minor success. This rather remarkable achievement in the fourth decade of Józef's life was all the more amazing because this budding young author was a relative newcomer to the English language—indeed, Polish-born Józef spoke almost no English before the age of 20.

The transformation of Józef Korzeniowski—who is better known in the English-speaking world as Joseph Conrad—from an adventurous man of the sea to a distinguished man of letters is one of innumerable examples of our rather astounding human capacity to change. After all, the operational abilities of Józef the Polish-French brawler, smuggler, ship's officer, and adventurer were transcended by Joseph the English aristocrat, intellectual, literary impressionist, and lecturer—widely identified as perhaps the greatest writer of his era—despite the fact that he was writing and speaking in his third language, working at his second profession, all of which he mastered as an adult.

Like Józef, you have the powers of transformation provided as a gift to you. How could you change your brain, if life circumstances demand it as they did for Józef—and in the end be a better, stronger, and even more interesting and more useful person?

The first great, paralyzing seizures began when Alex was in the fourth grade, but there had been a number of earlier signs that things might not be quite right in his brain. By the time he was a sixth grader, both small and large seizures were occurring with great regularity. No drug could control them. As Alex's life prospects grew darker, his father, in frustration, asked doctors at the famed Mayo Clinic in Rochester, Minnesota to determine what was really wrong with his son. Now in the hands of Mayo's experts, the diagnosis quickly came in. Thirteen-year-old Alex had a large oligodendrocytoma—an invasive tumor—in the left hemisphere of his brain.

For Alex and his parents, there was never any choice about what to do. A large part of his brain—the left temporal lobe and the underlying amygdala and hippocampus on the left side—had to be surgically removed to destroy the tumor and quiet his brain. The decision was easy to come to, but hard to live with, because everyone knew that voluntarily inflicting this level of brain damage would further degrade Alex's language, memory, reasoning, and social cognition abilities in ways that could be expected to reduce his chances for living a normal, socially effective, and successfully independent life. And how could we ever expect Alex to really recover, when the chunks of excised brain accounting for such important neurological functions would easily fill a soup bowl?

After the operation, Alex struggled in his social life and impulse control, and he suffered from cognitive control and memory association deficits that gravely impacted his academic and social accomplishments. Despite all of his challenges, he was able to educate himself as an artist, very capably developing right-brain skills and abilities that were not impacted by his left-brain tumor or its surgical removal. Still, his struggle to survive and thrive was frustrated by periods of depression, thoughts of suicide, intense crying fits, and by other problems in social and emotional control. His memory problems and his inability to sort out and remember the labels of the things of the world were limiting and frustrating. Alex was a poor reader. His handwriting was crude and halting. Ultimately, he developed more and more problems controlling his mood, his social ineptitude frustrated the achievement of personal happiness, and he felt increasingly disconnected and unstable in his thoughts.

"Going backward" wasn't exactly the path to a fulfilled life that Alex was seeking! At the peak of his frustration, he asked himself how he might turn his life around, to go forward once again.

As he approached his 30th birthday, Alex committed to take more advantage of his brain plasticity—with the fuller understanding that he had, within him, significant powers for transformation. Alex now works for about an hour most days at the brain fitness center at the BrainHQ website. In every way that he can assess, he's transformed his brain and changed his life. He has recovered the ability to remember the important things that happen in his life. His immediate memory abilities are reinforced by an ability to better categorize and label the things of the world, to keep things straight, and to grow the associations in memory that contribute to greater understanding. Alex's reading has greatly improved, and he now reads voraciously. He began to write down his thoughts in a diary at about the time he initiated serious training. His jottings were brief at first, his handwriting crude, his written expressions limited. Two years later, he writes rapidly, fluently, and with a good hand.

A drawing Alex sent me after our first meeting.

Alex has now shifted his perspective from a predominantly negative and pessimistic one to an appreciation and focus on the brighter things of life. His recovered optimism is supported by improved social abilities, guided by making better judgments about where other people are actually coming from. Alex now more appropriately responds to other people socially, controlling his impulses and inhibitions. He is a delightful, fluent conversationalist. Life is happier. With the clear understanding that his transformation to a stronger, more useful, and happier life is still a work in progress, Alex has recovered his confidence and his joie de vivre.

You also have the powers of very substantial positive improvement—and very possibly, to a truly transforming recovery after damage to your brain—if, like Alex, that is one of your burdens in life.

Nancy was 67 years old when she received her first diagnosis of cancer. As a young woman, she had been a fashion model and secretary. She had married a doctor and dedicated herself to rearing her three young children. Although she was socially active and had worked for many years as a volunteer at a local animal shelter, she was never officially employed after her children left home.

As with many other sufferers, the diagnosis of cancer led Nancy to reflect on how she had spent her life. Had she followed her dreams? Beyond her beloved children and family, what had she contributed to the world that anyone would remember after she was gone? She had always loved cooking, but had never pursued this beloved hobby to any great end. She enjoyed music in almost every form, but beyond the annual Christmas carols or neighborhood sing-alongs, she had developed no special musical ability. She thoroughly enjoyed arts and crafts and had been doing a little casual painting over many years in her life, but had never developed any particular skills in art, sculpture, sewing, or ceramics.

"I've always loved representational art," she told herself. "Maybe it is not too late to really learn how to paint."

My wife and I have five of Nancy's paintings hanging in the rooms of our home. This dear lady was only given four more years in life to develop and refine what turned out to be truly wonderful gifts of aesthetics and manual skill. We do not hang these pictures on our walls merely because we were very fond of Nancy. We love them. They represent what we see, at the best moments and in the best light, in our countryside and in our life.

Nancy transformed herself near the end of life. While she struggled with her illness and the tortures of being treated for it, the truth is that she may never have been happier in life then in her last four years, when she found new skills and abilities at which she truly excelled. She managed to leave a material mark behind after she left us. We visit it every day. My wife Diane and I often ponder, "What if Nancy had undergone this transformation 10 or 20 or 50 years before? Would almost everyone know about her wonderful ability to represent or evoke deep thoughts about the things of the world with paint?"

There are things that are left unfulfilled, unmastered, unlearned in almost every adult life. As we grow older, we have an increasingly clear understanding of what we would really enjoy spending more time doing. What things in your life remain unexplored and undone? Your plastic brain is waiting for you to take greater advantage of your potential for further personal development.

Chester was a hack, elevated to a lucrative position simply because he so reliably obeyed every order from the puppet-master who pulled his strings. It was somewhat embarrassing when he was put forth as a candidate for high office, since there had been almost no useful record of work on his resume. Unfortunately, the appointment of an amoral flunky was a critical step for getting his master—the most powerful politician in his political party in his era—to step aside. His political colleagues grudgingly accepted his nomination to assure that a man of exceptionally high purpose, ability, and integrity was nominated into the top slot. "Not to worry," they told themselves. "Our man James is young, vigorous, at the peak of health. There is no real chance that the worthless use of skin that they call Chester will ever be in charge." The gods favored their chosen candidate. James was elected to the highest office in the land without ever leaving his front porch, without having to raise a single dollar to mount a winning campaign against a worthy opponent.

How could they have predicted the assassin's bullet? How could they have imagined that the brilliant reformer would leave them not many weeks after assuming executive control? U.S. President James Garfield had credentials for changing the government and country for the better that were probably unmatched since the time of Abraham Lincoln. Because of a crazy man's bizarre personal vendetta, James Garfield was gone and "Mr. Do-Nothing" Chester Arthur now held the highest office in the land.

Much to everyone's surprise, President Arthur rose to the occasion.

Coming out of his daze, he pushed strongly for the very reforms in government that had been the cornerstone of Garfield's program—all stridently opposed by the powerful manipulators in Arthur's own political family. Historians now recognize him as a man who intelligently and forcefully governed the nation with moral authority for the greater good, to the benefit of all concerned. Chester transformed himself morally, ethically, spiritually, in the seventh decade of what had been, up to that point, a largely wasted younger life.

We all have it within us to rise to a higher plane. We have the capacity, as long as we're alive, to change the higher operational principles that guide our very soul—to change the person that we are, for the better. You share that great gift with Chester, and with the many millions of other individuals who have carried themselves to a higher plane when they squarely faced the bald realities of their own lives and its challenges. You and I are always subject to social and spiritual improvement. What, in the end, could be of greater true value to us?

For further explanations and extensive references and citations
related to the information in this chapter, please visit
www.soft-wired.com/ref/ch07

PART TWO:
BRAIN PLASTICITY THROUGHOUT LIFE

8

THE HUMAN ROLLER COASTER

The Rise and Fall of Human Ability Across a Lifetime

Our brains begin to change plastically when they are first bombarded by information received from our senses, which happens at roughly the beginning of the seventh month of fetal life. Of course, the world that the fetus is trying to understand is a dark and underwater place where sound is muffled, movements are constrained, there's a lot of bouncing around in unpredictable ways, and there is no voluntary action in play. Needless to say, those earliest brain changes are expressed in only very crude and limited forms. We understand how limited they have been when the baby arrives in the real world with only primitive capabilities in perceptual, cognitive, movement, social, and body function control.

The connections between the major parts of an infant brain are established on the basis of genetics. On the other hand, the specific detailed brain connections that will account for almost all organized and refined behaviors of the person to come are—of course—not yet in place. At birth, the machinery of the brain is very noisy, very disorganized, very imprecise, very slow in its operations, and very, very, very unreliable. It's no accident that we cannot really remember very much if anything of what happened in our early childhood. Things are amazingly noisy and chaotic in the fetal and newborn infant brain.

Scientists have documented a long list of characteristics in these young brains that account for their limited performance abilities. Just about every aspect of their neurology is unrefined. Brain processes are sluggish. The brain can only crudely reconstruct the details of what the baby can see or hear or feel or smell. Baby brains cannot accurately respond to or interpret the fast-changing, rapidly successive visual, auditory, or body movement events that we absolutely have to master to operate with any proficiency in our real world. The infant's brain machinery is susceptible to even the smallest of distractions. The interconnections of the wiring of brain cells with other nearby neurons—reflecting the extents to which brain cells cooperate with one another—are disorganized, weak, and diffuse. The flip side of this brain cell cooperation is uncertainty and unreliability. Because of this neurological chaos, and because the wiring of the brain is very poorly insulated, the higher levels of the brain receive relatively little interpretable information about what the baby is hearing, feeling, or seeing.

We can list several thousand molecular, cellular, and general structural distinctions that mark a newborn's brain, distinguishing it from an older brain. As a brain advances in its capabilities from birth across childhood and into adulthood, all of those chemical, structural, and functional attributes must (and do) change.

Scientists have long described these age-related changes in our very young lives as representing the final stages of brain "maturation" or "development." They previously regarded this "post-natal maturation" as the end-stage of our human embryology. We now know that the development of a "more mature" neurology is staged in your brain, and that by that interpretation this hypothetical late embryology actually extends out about two decades after our arrival in the world. Brain areas that most directly receive high-quality information from our senses "mature" first. As they are modified by our experiences to become more reliable, refined, agile, and organized in their operations, they deliver better-quality information to the higher brain regions that they feed—which, in turn, enables their "maturation." Because there are multiple stages of higher and higher levels of brain processing in humans, the progressive "maturation" of those successive stages contributing to our progressively more powerful cognitive competencies normally takes about 15 to 25 years to reach their peak. This maturation occurs in the cerebral cortex, which is the brain structure responsible for most of our remembering, reasoning, predicting, action planning, and thinking.

It is very easy to change the speed at which individual brains grow up. In experiments, we have arrested brain development for the listening brain almost completely, for example, by never allowing a baby rat to hear any sounds without also hearing moderately loud hissing noises. The development of refined processes of the brain—those several thousand attributes that we associate with its "maturation"—are nipped in the bud by just never allowing that growing rat to hear clear (noiseless) sounds as they advance toward physical adulthood. Put

41

another way, without normal hearing experiences, none of these neurological processes advance in the hearing brain in the normal way because progress is critically dependent upon and controlled by plastic changes driven by that animal's (or human's) listening history. Such a rat, impoverished for meaningful sounds, has the physical body of an adult, but their auditory brain still looks very much like the ill-formed, chaotic brain of a baby!

We can also just as easily speed up these "maturational" processes. Any brain, at any age, can be made to operate more like the refined machinery that is in place in the peak-performance epoch of life. In humans, our brains operate with greatest accuracy, reliability, and efficiency, on the average, in the third decade of life. If I were to train a child or an older adult in appropriate ways, I could easily drive many of those several thousands of elemental processes that define the state of brain maturation in ways that make you operate more like a 20-something! Such improvements can be achieved for almost any younger-than-20-something—or older-than-20-something—individual.

If you are an English major, you might be getting more than a little annoyed that I have put all of those quotation marks around the words "mature," "maturation," and "development." I have done that deliberately because those several thousand cellular, molecular, chemical, and structural processes that have been thought to advance as a final extension and completion of a mammal's embryology are actually a simple result of reversible brain plasticity. All of these aspects of our neurology advance in a normal individual as their brain ages into their 20s. Then, on the statistical average, as the years pass by decade by decade, these very same processes ALL roll backward, slowly, in the infant-ward direction.

We discovered this important fact about the older brain by documenting a long list of chemical, structural, and functional brain features in animals near the end of their natural lives, comparing them with animals in the very prime of life. Every feature that we studied was degraded in the old brain when compared with the vigorous young-adult brain. It took us awhile to understand that the lousy status of the very old brain actually largely duplicated the identically lousy physical, chemical, and structural status of a baby brain.

To further demonstrate the fundamental reversibility of plasticity-driven processes in the adult brain, we then trained old rats whose brains were in a grossly deteriorated state as they were about to die of old age in ways designed to drive brain processes in a youthful direction. By the use of only two intensive training strategies, every functional, chemical and physical feature of the brain that we examined was transformed—reversed—from the aged-rat form, back sharply in the direction of or completely back to the peak-performing young-rat form. We also brought rats in the prime of life into continuously noisy environments. Within a few weeks, Mr. and Mrs. Rats' hearing brains looked like they were near the end of their days on earth—and at the same time, looked just like the hearing brain of a rat that had almost no earthly experiences—because chemically, structurally, and functionally, very old brains are very much like very young (baby) brains.

From your brain's perspective, its life has been a kind of roller coaster ride, with one very large hill on its path. That hill is slowly climbed over the first two or three decades of life. The average modern human reaches the peak of neurological performance characteristics in their 20s or 30s, when many of us (me included!) are too young and dumb to fully exploit them! Then, most of us slowly decline. In a sense, our machinery runs backward over those five or six final decades in life in the direction that it came from in earlier life. In the beginning, our brains were noisy and chaotically imprecise; as we age, they slowly roll back in the direction of noisy and chaotic imprecision. Along this path are many smaller mountains of achievement and dips in ability, as our reversible plasticity processes empower or degrade our abilities day by day, week by week, and year by year.

The very good news is that if you've never quite grown in a way that has carried you to an especially high performance level, you can still grow that peak, at any time in life. Even if you've been sliding backwards as you move past (even far past) that peak, given its reversible nature, your brain plasticity is still there to call on, to engage to help you recover what can be, in a sense, a far more youthful stature. And if you've been fortunate enough to have sustained yourself at a high performance level at whatever age, you have the rather remarkable power to grow your mountaintop. It's very likely that things can look still brighter from up there!

For further explanations and extensive references and citations related to the information in this chapter, please visit
www.soft-wired.com/ref/ch08

9

CRANKING UP THE NEW MACHINERY

The Acquisition of Skills and Abilities in Early Life

As we've just discussed, despite all that they have to learn over a course of a modern life, babies—even those in the 21st century—are born stupid. Beyond crying, sucking, and pooping, newborn infants have little ability to control their movements. When they first emerge into the world of bright lights and clear sounds, they have only crude receptive abilities. And while they've bounced around a lot in the womb, they have a very primitive sense of balance, and about as much postural control as a boneless chicken. There's no evidence of thinking or reasoning ability, little sign of a personality, little evidence that there is actually a "person" on board, making decisions and directing brain traffic.

Now jump forward and think about the elaborate functional capabilities of a 10-year-old child. I have my wonderful granddaughters Leila and Mitra in mind as I write this. From their first wiggles as babies through rolling over, sitting up, reaching, grasping, crawling, standing, walking, running, and jumping, 10-year-olds can control the complex movements required to ride a bike, play a complex game like soccer or baseball, control a keyboard or smartphone like a wizard, play the piano, or whistle a merry tune. From the first gurgling sounds, Leila and Mitra have graduated to talking a mile a minute. Leila is a good soul with a great inventive spirit—for example, all on her own, as a gift of kindness, she wrote a beautiful song that she sang in her wonderful lilting child voice at my youngest daughter Karen's wedding—making everyone there smile and cry. Mitra loves to draw her grandpa into her games and puzzles so she can tease him and show him, once again, that her reasoning and thinking abilities can top the old fellow's!

By age 10, each child has a clearly defined personality. If we have

lived with a child since birth, we have witnessed the slow, progressive emergence of a unique individual, special and different in behavior and perspective, distinguishable from any other in the world. This truly remarkable development
of skills, knowledge, and "personality" from birth to age 10 results from massive physical, chemical, and functional changes in the child's brain.

Understanding the nature of the acquisition of our skill repertoires in young life provides an important foundation for understanding why our skills deteriorate in older age, as well as insights into how to prevent or reverse that deterioration. That acquisition is achieved across two great, complexly overlapping childhood stages:

Stage 1: Brain machinery setup in the "critical period." In infancy, and progressively staged across those higher and higher brain levels through childhood, plasticity goes through a critical period, a time of riotous brain change. Through the critical period, each functional zone of the brain is remodeled to make its own special contribution to that long, slow process of creating an effective, operational person from the tabula rasa (the "blank slate") that begins to organize itself functionally, in utero, at about the beginning of the third trimester.

You might think of the newborn brain like a marvelous factory whose owners still haven't decided exactly what they're going to manufacture. The baby brain doesn't know precisely what tools and materials it needs to make its products. As a result, its plasticity in the critical period is unregulated; the "plasticity switch" is always "ON." Everything the baby sees or hears, every wiggle or babble it makes, drives plastic changes in Baby's brain. In this magical period of "anything-goes" plasticity, the brain literally takes in information like a sponge. She doesn't know—cannot know—exactly what is important in her world. For Baby, everything matters. Everything drives brain change.

What, exactly, do I mean by plastic brain changes? There are many components of change that we'll talk about in later chapters, but here are a few basics: When our bodies sense a sound, a feeling, a sight, or a smell, then our eyes or skin sense or ears or nose translates it into patterns of electrical impulses that engage the brain. Those electrical impulse patterns travel through the brain on incredibly thin "transmission wires" (axons), and are complexly conveyed in the brain from one brain cell to another. As a skill is developed (such as whistling, or doing a pirouette, or identifying bird calls) the specific neural routes that account for successfully performing this new skill become stronger, faster, more reliable, and much more specific to—specialized for—the task at hand.

At birth, the brain has inherited neural pathways that could be thought of like the main trunk lines of a great transportation grid. You can imagine the newborn infant's brain as like a highway map of North America or Europe or Asia with just the largest freeways and most important highways laid onto the map. Those major thruways interconnect regions to one another—but no one has yet constructed any local highways, secondary roads, streets, byways, lanes, driveways, or garden paths. Most places (specific abilities) remain inaccessible until these routes are in place! In the critical period, the brain begins to build many of these routes.

As the brain's processing machinery begins to develop, it becomes specialized for recording "What's that?" and "What's happening?" in the environment into which the baby has been born. Its great goal is to create an effective, useful understanding of that world, and to learn to appropriately control its actions within it. The details of environment are very different for babies from different cultures and parts of the world. Consider the language that the child is born into as an example. The child that arrives in a San ("Bushman") family must quickly develop a language interpreter in its brain that is specialized for its native language sounds. Its brain must, soon afterward, learn to control the jaw, voice box, upper throat, tongue, and lips in the sophisticated ways that are required to produce the special sounds of the San language. The Japanese or Lett or German or Tagalog or English or any-other-native-language kid's environmental exposure leads to the emergence of the appropriate language-specific brain machinery that enables language development in that child's home. Baby (her brain) has nothing to say about her fate in this respect. If the only sounds that Baby Wolf-Boy ever hears are the cries of the wolf, he will become a master receiver and producer of the barks, chortles, and yips of Wolfese! The brain and its neural networks actually change—through brain plasticity—as the sounds are received. By those changes, the special machinery that is designed to sort out and ultimately interpret just those sounds of Baby's native language emerges in the brain. In the same way, the infant brain undergoes a myriad of environmentally specific changes beyond language that organize its evolving "factory" to get the most out of the massive amounts of information streaming in from the specific world into which it was born.

Across this early critical period, the brain also gradually learns to control brain change. It has almost no such ability in early infancy; for Baby, plasticity is unregulated, and almost any experience will change the brain. After all, when Baby was born, its brain had very little ability to reliably or selectively encode and record the things of its world. That rapid creation of special neurological constructions ("representations") of a million and one things of the baby's environment is an absolutely essential first step for the brain's emergent control of selective attention. A brain that has little ability to be selective (and thereby, pay attention to just this, while it ignores all of that) in its neurological responses to what it sees or hears or feels in the world is not going to be able to control and focus its learning in ways that shall be required to master skill after skill after skill in later childhood.

In the older brain, most experiences won't result in enduring brain change. In the older child and adult, in most brain regions, the plasticity "switch" is mostly turned "OFF." Change is only permitted for those things that have captured the brain's attention, and only when the brain itself judges that change to be beneficial for it. This crucial control capability also emerges progressively, through plastic changes in the brain, during the critical period.

What brings this early "anything goes," "always-on" critical period era of plasticity to a close? The brain is engaged by the repetitive sounds or feelings or sights or smells or whatever fills its days, and it is engaged by every little movement and "ba-ba" and "ma-ma" that the baby generates. Its plasticity processes competitively sort out the responses evoked by specific sounds or feelings or sights or wiggles or babbles, and those millions of little moments of Baby's special environmental experiences come to more and more selectively and more and more strongly activate her brain. Ultimately, neurons in this competitively plastic machinery create a reliable, sorted representation of the details of that very special world of Baby. That growing reliability is detected by the brain. It informs the brain that the factory's machinery is now finally ready for more controlled action! Given the very good start that a normal critical period can provide, the brain can now develop the million and one skills and abilities that the brain itself determines are in its own best interests.

Stage 2: Brain plasticity in the older child and adult. From the end of the critical period forward to the end of life, the brain controls its own self-development—its plasticity. It no longer absorbs everything it hears, sees, and feels. This is a good thing. If the adult brain made enduring plastic changes based on everything it sensed, certain experiences would become grossly and highly inappropriately overrepresented in the brain. As an example, since most of us spend a lot of time sitting down, an enormous area in our brains would become dedicated to information from our fannies, which isn't exactly the best possible use of our brain resources!

How do the changes in the brain convert it from an anything-goes machine to one that has control over its plasticity? The biological principal is not complicated. In the beginning of life, the "plasticity switch" is always turned "ON." As the brain matures (generates more reliable and coordinated responses), it undergoes physical and chemical changes that increase the power of the "OFF" switch. Over time, the balance of power changes. The "OFF" switch dominates, and plasticity is only flipped to "ON"—permanent changes in the brain's machinery are only permitted—under certain circumstances:

- when you pay careful attention, or focus on a task or goal;
- when you (your brain) are (is) rewarded or punished—or expect(s) a reward or punishment;
- when your brain positively evaluates your performance in a goal-directed behavior;
- when your brain is surprised by—or potentially threatened by—something new or unexpected.

In other words, the older brain allows change to occur specifically when it determines, by its own standards, that change would be good for it!

Now that the older brain has finally established control of its plasticity switch, it can throw it in the "ON" or "OFF" direction by controlling the release of small chemical molecules called "modulatory neurotransmitters." For instance, if you achieve a learning goal or feel successful, one of those modulatory neurotransmitters, dopamine, is released to tell the brain to "Save that one!" If you're positively or negatively surprised, another neuromodulator, noradrenaline, pumps out to tell the brain "Wow!" or "Watch out!" If you're totally absorbed by a task, acetylcholine spritzing tells the brain to "Try this one!" as it amplifies important new options for changing the brain, so that it can improve at the task. The flip side is that when task achievement is of no consequence, or when experiences are unimportant or strictly routine, the brain does not release these chemicals. The plasticity switch remains in the "OFF" position—and nothing changes in the brain.

It is easy to demonstrate the basic plasticity of the older brain. In thousands of studies, scientists have documented the neurological consequences of teaching an animal or human child or adult new skills and abilities. Teach a monkey to recognize colored visual shapes that they could never have seen before, and you can literally witness neurons becoming specialized for responding to just those shapes, while neurons in the brains of their non-exposed buddies show little interest in them. Keep up the training for a year and see hundreds of millions of neurons develop a strong, special interest in just those strange colored shapes.

Teach a kid or adult to play the violin, and after all the squeaks and squawks, the brain regions for hearing, fingering, and reading music all specialize to manifest that growing mastery of tone, timbre, precise timing, complex sound combination, finger pressure, musical note-to-sound equivalences, sound sequencing, and movement translation from a musical score. Their brain has created a music-mastery brain system that my brain—and unless you've been seriously trained in music, your brain—does not have.

Changes in the brain resulting from learning any new skill are massive. An example of a skill that we've all mastered is the development of the facile control of a simple manual feeding tool, the spoon. I've had great fun watching my own grandchildren slowly acquire mastery of this tool. Its effective use takes many thousands of practice tries extending over two or three years of a young child's life. What has to change, in the brain's machinery, to perfect such an ability? First, practice with a spoon improves the quality of information that the child's brain receives from the skin of the hands, and from muscles and joints in the fingers, palm, wrist, and arm, because the child has to successfully support a spoon carrying different quantities of food with highly variable properties. Very sophisticated sensory feedback-guided adjustments in posture and fine movement are necessary because of variations in the spoon, in where the child happens to set the fulcrum of this lever when it is grasped, and in the viscosity, specific gravity, weight, and texture of the food loaded onto the spoon. Is it grits, granola, gravy, grapefruit, or gummy bears? Each requires different, subtle, rapid postural adjustments to load the spoon neatly, bring it to the mouth, and reliably dump its load with stable and fluid movement.

To accomplish and coordinate these changes, the brain had to remodel its neurological coding and representations of sensory information from the skin, muscles, joints, taste buds, nose, and eyes, and must modify the machinery that underlies planning for, and the production of, incredibly sophisticated movement sequences that are guided by target identification and very complex moment-by-moment feedback. This specialization is achieved through changes in the responses of hundreds of millions of nerve cells, and of <u>billions</u> of connections between them.

What drives all of this behavioral improvement? First, the food itself—which is the child's "reward" that tells the brain to "save" the activation patterns from each attempt that results in the child receiving that very tasty noodle or dumpling; and second, the child's brain's own assessment of the accuracy and effectiveness of the try. It knows, from watching and doing, what good spoon use is all about! It knows (and rewards itself) when it has done just a little better job of making a "good try."

If this massive brain alteration is required to use a spoon, just imagine what the brain has to do to master our great skill repertoires—language, reading, computer programming, mathematics, sports, art, music, and so forth.

It is important to understand that even the operations that control your learning ability and style are plastic. One of the first things that the brain has to achieve is to learn how to learn. "Gee," you're protesting, "surely I inherit that." In fact, specific learning strategies are largely acquired in your brain as early habits. Some of those habits develop through the progressive natural evolution of each baby's behavior. At the same time, learning strategies are reinforced by the teaching strategies of a child's parents and mentors, or by the difference in mentorship that is a natural result of being in a different sibling position in a family. For example, one child might "instinctively" endlessly identify and label the things of the world; a second might explore the relationships between the things as a world; a third might exaggerate an interest in how the action of any one thing affects others. One child might emphasize the visual; the second might more strongly emphasize listening; the third (like my grandson Gus) might be "all action." A typical American parent, on the statistical average, places a stronger early emphasis in teaching an infant and toddler the names of all of the things of the world. "What's dat?" asks Sissy. "That's an aaapppplle," says dad. A typical Japanese parent might place more emphasis on the relationships of the things of the world, and respond: "Sore wa ki ni naru kudomono desu." (That is a fruit that grows on a tree.)

Each of us develops our own special learning strategies. A scientist can see them in operation in simple learning that reflects how the brain prepares itself to receive information, and how it uses it once it arrives. These different learning strategies substantially account for the theorists, inventors, engineers, builders, artists, accountants, teachers—and many other common (variously successful/unsuccessful) operational styles of learning and skill development seen in our species.

We shall later consider how understanding and exercising your learning strategies—or if necessary, plastically re-shaping them to gain greater learning power—is a core aspect of growing and sustaining your own brain health.

I also want you to remember that the course of normal development of the abilities of any human individual is literally a unique and substantially unplanned childhood adventure for every one of us humans. Almost every Inuit boy along the Northern Alaskan and Canadian and Russian coastlines knows how to skin and flense a seal. You don't. About half of the boys in Sao Paulo know how to bounce a soccer ball on their head while running full speed. I think that I'm safe in guessing that you do not. Nearly all girls in Palau talk with gestures while they dance. You probably can't do that very well. Iroquoian children from several hundred years past all understood that you honored your enemy by removing their beating hearts and taking a bite out of them. You did not learn this. A hundred and more years past, most sophisticated Chinese women bound their feet so that they slowly resembled unfolded lily blossoms. It's unlikely you did that.

There are a million variations of culture that are widely expressed in some but not other societies. The specific skills and social principles that we acquire specializes us (our brains) for our local societies. In many thousands of other subcultures, we would be largely "out to lunch" brain-wise, for much of every day! As I have repeatedly pointed out, the accomplishments of brain plasticity are remarkable for the fact that they load each one of us chock full of skills and abilities and information and concepts that distinguish us, in detail, from every other human being that has ever lived—or ever shall live—on the planet. At the same time, from the perspective of the zillions of culturally specific abilities that humans have mastered across our species' history, we are, individually, very small cheese! While our brains—and all other humans' brains—are literally seething with change across childhood and into young adulthood, the ultimate operational products of all that plasticity are rather remarkably variable! In organizing a personal plan to determine what you might need to do to sustain and grow your special abilities, a thoughtful self-assessment (and possibly, a formal clinical assessment) of that very complicated brain-plasticity creation that you are—with all of its strengths and weaknesses—is a key starting point.

Of course neurological progressions can also be altered in small or in large ways by differences in our genetic inheritance, but those genetic variations must still be played out in a highly plastic brain. An illustrative example is provided by the development of language in a child born with a deep cleft palate, born into a home where the native language is English or Swahili or Arabic. Because this inherited fault splits the palate on the roof of the mouth, it closes off the two long tubes that drain fluids between the eardrum and the inner ears. As a result, all hearing in the cleft palate child is essentially "under water." From the brain's perspective, with that cleft palate, the child's native language is muffled English or Swahili or Arabic.

Until about 40 years ago, scientists and doctors generally agreed that the genetic fault that caused the cleft palate also caused grossly impaired language and speech abilities, and a substantial degree of "mental retardation." All of these problems miraculously disappeared when it was discovered that none of these negative outcomes occurred if the cleft palate was surgically repaired before age one. With the repair of the palate, the fluid drains out from behind the eardrums as in any normal kid. In a flash, the native language model of the child now has normal power and clarity. Surgically corrected cleft-palate children who are still in the critical period can still use this now-clear language model to re-specialize their brain machinery to accurately represent all of their language's rich features and complexities. Cognitive deficits were not a genetic part of the cleft palate syndrome after all—as had so long been believed.

Of course most genetically imposed limitations, or strategies by which we might hope to overcome them, are not so obvious to us. Still, almost every living human has a plastic brain, capable of further empowering or correcting its impaired operations. Neither children nor adults are stuck, brain-wise, if they just happened to have been born into a family with less-than-perfect genetic attributes—if there even is such a thing (for heaven's sake!) as a perfect genetic background. Certainly my family doesn't fully qualify as genetically perfect! How about yours?

Don't lose sleep over it. Your brain is prepared to help.

For further explanations and extensive references and citations related to the information in this chapter, please visit
www.soft-wired.com/ref/ch09

10

HOW DOES A BRAIN REMODEL ITSELF?

Ten Fundamentals of Brain Plasticity

I would like to pause in this narrative to briefly summarize the rules that govern the brain's self-organizing plasticity processes. It is obviously important for you to understand just what your brain is up to when it changes itself, for what it judges to be in its best interests. To briefly review the core principles in play for the plastic remodeling of your brain:

1. Change is mostly limited to those situations in which the brain is in the mood for it. If I am alert, on the ball, engaged, motivated, ready for action—the brain releases those chemical modulatory neurotransmitters that enable brain change. Again, it's helpful to think of them as on/off switches. When I'm in a learning mode—alert, concentrated, and focused—the brain's plasticity switches are turned "on" and ready to facilitate change. If I'm disengaged, inattentive, distracted, sleeping, twiddling my thumbs, doing something without thinking about it, or performing an action that requires no real effort to succeed on my part, my switches are mostly turned "off."

2. The harder we try, the more we are motivated, the more alert we are, and the better (or worse) the outcome, the bigger the brain change. The machinery that enables brain change has a dial on it that can mean "ready" or "<u>ready</u>" and "save it" or "<u>save it</u>." If you are paying just a little bit of attention, are half-trying, do just a tiny bit better than the last time, or receive a penny for your success, then only a small dose of modulatory neurotransmitters are released and the attempt results in only very small and ephemeral changes. On the other hand, if you're really intensely focused on the task, are trying as hard as you can to get it right, do it a lot better than on your last attempt, and receive $100 for your success, neurotransmitter release rises dramatically—and large-scale and long-enduring changes will result.

3. What actually changes in the brain are the strengths of the connections of neurons that are engaged together, moment by moment, in time. Your brain's primary trick is to select all of those activities that contribute to a successful behavioral try, for each important moment in time during that attempt. It does that by simply making the connections between brain cells (neurons) for all of the simultaneous momentary activities contributing to a little more success just a little bit stronger. For example, if the brain's job is to progressively establish reliable control of peeling and eating a banana, it achieves that by strengthening all of the lines of information that are nearly simultaneously engaged, at each brief moment, for the acts of grasping, orienting, opening the top, peeling back, and consuming that banana. As I practice that ability:

a. I co-strengthen the specific activities that provide all of the information that applies for just those things that occur nearly simultaneously, moment by moment in time, that represent banana-ness:

i. the sensory information representing the size, shape, texture, color, and aroma;
ii. the cognitive information that represents the eatability, the anticipated pleasure of eating, the quality of the banana being eaten;
iii. the sensory information and motor control of banana handling—contacting and grasping, orienting, breaking apart the skin segments, releasing the aromatic blast, peeling them back, bringing the exposed fruit up to the mouth, taking the bite;

b. In the background, I am refining my brain machinery by strengthening the connections of all of those sources of information that arrive nearly simultaneously, moment by moment in time, that represent the banana-specific details of their smell and taste and mouth-feel; the sensory information related to the complex sequences of postural control, grasping, peeling, and consuming the banana; and a cognitive level that is all about remembering past bananas, predicting every little step in the handling, eating, smelling, consuming, banana-pondering and umpty-ump other banana-related processes; how much I like bananas and how this banana measures up...and many thousand of other possible ways in which learning to eat and eating all of those bananas result in brain rewiring!

Go through these rewiring change cycles a large number of times (we call that "practice"), inducing changes in millions or billions of nerve cell-to-nerve cell connections, and your brain has created a "master controller" that can implement this or any other practiced behavior with astounding facility and reliability.

4. Learning-driven changes in connections increase cell-to-cell cooperation, which is crucial for increasing reliability. Imagine that you are sitting in a large football stadium, and that all the fans in the stadium are clapping at random. A low dull roar rises from the stadium. Now imagine that 100 of those fans begin to clap out a rhythm in unison. Rather remarkably, you can hear that rhythm as a distinct "signal" that rises above that background noise. An analogous growth in brain cell coordination is exactly how your brain is generating representations of the things and relationships and actions of the world.

As you are listening to those people clapping in unison, imagine how the clarity of that rhythm would grow if they were even more precisely coordinated. Or imagine how the clarity would grow if the numbers of simultaneous clappers slowly increased. Again, the recruitment of "team members" and a steady increase in the coordination of their actions is exactly what your brain is up to when it's in a learning mode. As with those clappers, more nerve cell team members acting in greater unison equates with growing power and reliability in how that message can rise above the din.

Your brain achieves this team-building by increasing the strengths of its interconnections between simultaneously excited nerve cell neighbors. In a well-organized brain, there are hundreds of millions of these nerve cell teams all fighting with their neighbors for the domination of every little part of your cerebral cortex. The more powerfully coordinated your teams are, the more powerful and more reliable their behavioral productions.

5. The brain also strengthens its connections between those teams of neurons representing separate moments of activity that

represent each little part of an action or thought. Things that we see or hear usually have many complexly related parts—for example, the successive sound parts of words in the syllable, the syllables in a word, the words in a phrase, the phrases in a sentence, or the sentences in a narrative. The brain strengthens its connections between its neurological representations of successive things that reliably occur in serial time. These linkages are critical for the brain's ability to predict and control the flow of all of your perceptions, thoughts, and actions. Without this continuous associative flow, each word and momentary thought and every other action would drop into the abyss; your glorious "stream of consciousness" would be reduced to a series of separate, stagnating puddles.

6. Initial changes are just temporary. They only become permanent if the brain judges the experience to be inherently fascinating or novel, or if the behavioral outcome is a good (or bad) one. The brain has the remarkable ability to first record the change, then make a determination—after the fact—of whether it should make that change a part of the permanent record. It does this by storing the change temporarily, then releasing modulatory neurotransmitters (brain chemicals) as soon as it is reasonably certain that the behavior has, or is likely to have, a good (or bad) outcome. The release of these chemicals turns the brain plasticity switch "ON," which converts the temporary plastic change into a permanent, enduring, physical change. During a success or reward, this chemical spritzing also induces the sense of pleasure or joy—or given failure or punishment, the displeasure or sadness.

7. The brain is changed by internal mental rehearsal in the same ways, and involving precisely the same processes, that control changes achieved through interactions with the external world. You don't have to move an inch to drive positive plastic change in your brain. Your internal representations of things recalled from memory work just fine for progressive brain plasticity-based learning! All of the other principles outlined in this chapter apply for mental practice, just as they do for improving your operations as a receiver of information from—or as an actor in—that external physical world.

To illustrate this point, I will share an anecdote. Several years ago, I was invited to speak with his Holiness, the Dalai Lama, about how brain plasticity science relates to Buddhist teachings. I found him to be one of the most exceptionally intelligent and inspiring individuals that I have met in life. When I introduced humor into the conversation, I saw instant joy on his face. When I explained how our science was targeted to help individuals in need, I saw instant deep compassion on his face. Most people do not have the ability to control their emotional states in this very remarkable way. It is, without doubt, a product of powerful brain plasticity generated by intensive, focused mental practice.

The Dalai Lama has been trained to use his mind for nearly 80 years, since the age of two. The Buddhist leaders identified him as the next Dalai Lama and began training him to use his mind—his capacity for internal mental exercise—to create this truly extraordinary and remarkably specialized human individual. His powers richly affirm many contemporary scientific studies that show that mental rehearsal induces these changes in the brain just as physical actions do.

8. Memory guides and controls most learning. Remember when you first learned to use chopsticks? You held these long sticks in your hand. You understood your goal, which was to deliver food to your mouth. You made a clumsy try. No food got anywhere, except on the tablecloth. Your brain remembered the goal, which you didn't come close to achieving. It said "Don't save that try." You made some adjustments for the second attempt, and some food—the reward—did reach the mouth. Your brain interpreted this try as positive when it referenced it to that remembered goal; it also referenced its own complex models of good chopstick use, which it had earlier established by observing proficient users, and from your trying it yourself. At this marginally successful try, the brain told itself: "Save this one." Progressive change required all that remembering. With another several thousand or two attempts, with continuous reference to your remembered goals and your models of good chopstick use, you can expect to become a chopstick master. This is equally true of all skill learning: without the remembered goals and strategy models that are required to define and guide positive incremental progress, you cannot progressively improve—or ultimately master—any new skill.

9. Every moment of learning provides a moment of opportunity for the brain to stabilize—and reduce the disruptive power of— potentially interfering backgrounds or "noise." The change processes in your brain do a wonderful second thing every time they strengthen the connections in your brain to advance your mastery of any given skill. After each moment of synchronized activity that leads to connectional strengthening, the same machinery takes the next moment in time to weaken other connections, just a little.

The brain's goal is simple. Its positive, connection-strengthening plasticity is increasing the power of connections on and between all of the brain cells that fire together at each moment of time, burning in those changes only if their actions contribute to success. Its negative, connection-weakening plasticity is reducing the power of the connections coming into the machinery or from other neurons that did not fire at that important moment. After all, a lack of response at just the right moment in time was pretty good evidence that they had not contributed to your success! Positive and negative plasticity work in concert. Positive plastic brain changes work to create a brighter and sharper picture of what's happening. At the same time, negative plastic brain changes are erasing a little of that irrelevant and interfering haze or noise that frustrate the construction and recording of a clear picture.

10. Brain plasticity is a two-way street; it is just as easy to generate negative changes as it is to produce positive ones. There are winners and losers in the game of brain plasticity. It is almost as easy to drive changes that can impair one's memory or slow down one's mental or physical control as it is to improve one's memory, or speed up the brain's actions. As I'll discuss in detail later, many older individuals are absolute masters at driving their brain plasticity in the wrong direction!

Scientists on my research team have conducted a number of studies illustrating this dark side of brain plasticity. For example, about 20 years ago, we conducted experiments that showed that we could turn a person's hand into a useless claw by engaging that unfortunate individual in a particular form of training for an hour or two a day over a several-week-long period. In the same way, if you would like to volunteer for it, I could easily train you over a similar time period in ways that would utterly destroy your ability to follow a normal conversation, or read the rest of this book.

Why would we use plasticity to drive negative changes? It turns out that millions of people have lost the control of their hand in what is called a "focal hand dystonia," which occurs as a result of how they use their hands in their occupation, through exactly this same kind of negative plasticity learning scenario. Previously, when a keyboard or computer mouse user, a professional musician, or an assembly line worker catastrophically lost control of their hand movements, scientists and clinicians blamed their dysfunction on physical changes occurring within the offending hand, arm, or other body part. We now know that such problems initially arise from negative, learning-induced plastic changes in their brains. Once our physical therapist collaborators began to re-train brains—rather than try to address these individuals' problems as exclusively residing in their hands and arms—they found that it was relatively easy to correct this class of severely disabling conditions.

Now that you have these ten principles of plasticity under your belt, you can sort out how to apply them to guide your life and behavior, on the path to recovering or strengthening—and ultimately excelling at—those key abilities that are the fundamental bases of a happy and fulfilling life for you.

For further explanations and extensive references and citations
related to the information in this chapter, please visit
www.soft-wired.com/ref/ch10

11

REACHING THE MOUNTAINTOP

Your Brain, in Technicolor

How much do you have in common with Albert Einstein? Among other things that could apply—frequent bad hair days, pronounced eyebrow unruliness, useful work conducted in a patent office or university, pipe smoking, reading impairment, chubby cheeks, a cute dimple, a good understanding of photons, a well-above-average laugh— you probably reached the peak of your neurological performance abilities, as he did, in the third decade of life. To understand how the brain advances from that primitive baby state to the remarkably elaborate and cognitively powerful peak performance level, consider how I am writing this book. Think about how I acquired the ability to control my arms and hands so I could enter the abstract information flowing out of my brain, in the symbolic form of the letters, words, phrases, and extended text that I am producing on my keyboard, just so you can literally read my thoughts, and interpret and record them in your brain.

In the domain of movement, your brain began the process of plastic self-organization in the womb by creating orderly, mapped representations of sensory information received back from the limbs and trunk. The brain then systematically related that feedback to the different patterns of excitation of the muscles that moved your tiny little arms, legs, and other body parts. This was accomplished by the operation of competitive plasticity processes that sorted out these crucial relationships. As a baby, you had to learn how to move and what consequence each movement would have. But you weren't going to move anywhere until those sensation-to-movement-to-sensation relationships were reliably recorded, mapped, and stored in your brain!

Out of the womb, we are immediately subjected to the powerful forces of gravity; our tiny arms are now heavy. If you have ever observed a newborn baby, you know that they move their limbs wildly and randomly at first. Babies keep flailing until their brains reliably chart the relationships between that sensory information coming back from their now-heavy arms, as a function of the patterned activation of the muscles that control their movements. These real-world corrections are much more complicated, because those arm weights are (of course) dependent on posture, and on whether or not any part of the arms or hands are supported. Your brain has to sort out and create a systematic mapping of all of these complex relationships before it has any hope of bringing your arms and hands under any sophisticated level of voluntary control.

Something else very important happens when we leave the darkness of the womb: we can now see our arms and hands, as well as all those million-and-one things that we might reach out for and touch or grab or manipulate with our hands. That seeing enables our brains to use information from vision to add to the power of its understanding of how muscle stimulation relates to arm, hand, and finger positions in three-dimensional space. Your brain actually controls the positioning of your arm in that 3-D visual space by very systematically representing how excitation of your muscles alters your view of its consequences. Those views of specific goals in your grasping behaviors are also crucial for your brain's keeping track of the relationships between the necessary arm and hand postures that are required to reach and get a grip on that toy rattle or piece of candy. If you want to see that marvelous ability in action, just watch your hand as you grab that coffee cup or cell phone or the next dozen or two things that you reach out to grab or touch or manipulate with your hand, and you will see a kind of biological robot in action working at a level of sophistication that engineers can only dream about!

At this point, the baby brain has all the crucial resources in place necessary for the successful voluntary control of movement. Most babies begin by reaching out to grasp a toy or other object in front of them to bring it to their mouth. From this small beginning, they progressively refine the control of arm position and fingers. They follow a painfully slow progression of fumbling and dropped toys to learn to coordinate the movements of their two hands. They learn to pick up even the smallest of things under difficult circumstances, and to pick up a wide variety of objects of different shapes, sizes, and weights. They learn to move fast and with high accuracy to master games like Whack-A-Mole and Slap Jack, and to hit that ping pong ball. They learn to control the timing and the individual and sequenced tapping of their fingers as they master the clarinet or piccolo or keyboard or cell phone. They learn to play out complicated hand movements to hit the pigs' houses with their Angry Birds, to play a Mozart concerto from memory, to caress someone they love, to shape the pot they're forming on the pottery wheel, to play out in their minds (imagine!) how they would have to move their hands to tie that knot with their hands out of view—and to achieve any one of millions of other possible things that fully controllable hands and arms can do. This is how we all learned to use (our brain remodeled itself to enable the facile use of) our hands.

Still, all of this brain change gets you only part of the way toward the rather incredible human abilities represented by my being able to type out the words that represent all of the language sounds and meanings in this book. That requires that the brain make equally astounding progress in the domains of aural language, in the special sub-domain in which we integrate our hearing brain with our seeing brain that we call "reading," and in our equally elaborate abilities to operate in the domains of recognition, symbolic labeling, memory, and thought. The brain has to create a reliable representation of the sound parts of words as a basis for translating them for facile visual interpretation by writing. It has to advance through long progressions in language, vision, and related thought-control to evolve the remarkable ability of having those thoughts flow out of my brain letter by letter through my fingertips in that wonderful translated, symbolic thought form that you are now reading.

How did my brain accomplish this almost unbelievable progression in skill development, from the first fetal wiggles to pounding out that text translating my symbolic representations of what I am thinking, to controlling my high-speed sequential finger movements, and to confirming that I've just said the right thing by monitoring my typing with my highly evolved vision and thought-control machinery? It's all a product of progressive, staged brain plasticity that takes many years to complete.

The brain uses two simple tricks to control the creation of successively more sophisticated abilities, always constructed upon a platform of skills that have already been mastered. As you improve in any ability at one performance level, the brain processes involved with that ability grow in their precision and reliability. As the actions of neurons in the brain become more coordinated, the machinery of the brain detects that growing teamwork and produces chemicals that enable performance achievement at the next highest levels. That's a pretty cool thing, when you think about it. It means that every time a child masters a new skill or ability, that mastery directly enables the next level of skill development! Standing enables cruising. Cruising enables walking. Walking enables running. Running enables jumping over hurdles or through hoops. Riding a bike enables riding hands-free. Every time the child succeeds at one level of performance, the child's brain opens the gates that allow that kid to move ahead to the next level.

How does the brain do that? The brain monitors the extent to which behavioral improvements are increasing the levels of coordination in its own neural responses. As that coordination grows to a level reflecting performance reliability, chemicals are released that induce a physical thickening of the insulation on the brain wires (axons) coming from this now-more-coordinated machinery. Wires with better insulation ship out more-coordinated information at higher speed. Because brain plasticity processes are coincident-activity dependent, the more coordinated that information delivered to the next-higher process level, the greater its powers for enabling change. Mastery at each level enables the initiation of effective brain change that can now (only now) lead to mastery at the next, then the next level—and up and up and up our brains go, advancing at a truly remarkable speed, in the elaboration of our skills and abilities. Each new level of achievement is based directly on already mastered abilities. The brain is sensible enough to wait until those resources that are required to support its next level of success are already pretty well established before it gives the go-ahead signal!

You can witness how this process advances by measuring the speed of accurate responding as an individual progressively develops any specific skill or ability. The more advanced the brain, the faster it goes.

I earlier alluded to the second great trick that the brain uses, when I described how the language-related memories of a cochlear implant patient actually control the details of brain remodeling required to re-establish their new speech understanding. As you develop each higher level of ability, your brain further refines all of the machinery that feeds that higher ability in ways that can contribute to its success. It accomplishes this by delivering top-down feedback to enable changes in all of the lower levels of the brain system that have contributed to moments of higher achievement on the task the brain is busy at mastering.

You know from your own experiences that the brain does not limit its plasticity solely to those moments when it is guiding changes just on the basis of information coming from the external world. It begins to grow the domain of thought too, by manipulating the information that you have recorded in your mind, using exactly the same plastic brain machinery. The construction of the machinery that controls and elaborates your mental actions in thought and reason, and the growth and refinement of the machinery that can so rapidly and flexibly adjust this analytic machinery on the fly, are among the last to complete their maturation in the 20-something's brain.

How long does it take for an average brain to complete this process of growing up, to approach peak speeds and peak performance capabilities in its highest-level operations? For women, this process is largely completed by the age of 16 or 17. Using the levels of axon insulation thickness as a guide, we see that the brain of a late teenage female appears to have reached its final state of "maturation." That doesn't mean that a 17-year-old girl has quite reached her peak, because her brain still has quite a lot of work to do to refine and elaborate its operations at those highest levels. That takes roughly another ten years of hard work.

The males are behind the females pretty much all the way through their childhood. A man is usually between 18 and 20 years of age when the brain wires are insulated most thickly in the brain's areas of highest learning.

At your peak, your brain operates with very high accuracy. It is really fast in interpreting what's going on and in responding to it. It is capable of operating at high speed not just because its wiring is well-insulated and you have high-speed transmission lines, but because it also operates with high levels of coordination—which equates with high accuracy and reliability—at every brain level.

At your peak, if you have been lucky in your genetics and on the nurture side of your life, you are using your whole brain to the max. You have great powers of abstraction. You can immediately recall and reconstruct most of what has just happened. You are capable of synthesis, invention, and discovery. When we move beyond this peak, although we are still likely to be growing in understanding and wisdom, we are also slowly moving backwards in terms of our accuracy, speed, action control, memory, and manipulation of information in thought.

If you are younger, understanding your brain in Technicolor is all about creating a higher peak and keeping yourself at or near that lofty position far closer to the end of your physical life. If you're a little older, it's all about getting back to those halcyon days of a younger-performing brain, so that the benefits of accumulated knowledge and wisdom that are the great virtues of your older age can enable a richer, fuller life.

At any age, it's all about doing the right thing for your brain, so that brain accuracy, speed, and complex control continue to grow, so you can continue to ride that bicycle hands-free, throughout life.

For further explanations and extensive references and citations related to the information in this chapter, please visit www.soft-wired.com/ref/ch11

12

SLOWLY SLIDING BACK DOWN THE SLOPE

How Cognitive Abilities Decline in Aging—and Imperil Our Older Lives

What did the author E.B. White, the celebrated chef Joyce Chen, the former British Prime Minister Harold Wilson, the novelist Iris Murdoch, the boxing champion Sugar Ray Robinson, the movie director Otto Preminger, the mystery writer Ross McDonald, Senator Barry Goldwater, the actor Peter Falk, the composer Aaron Copeland, the artist Norman Rockwell, the actor Charles Bronson, the crooner Perry Como, the actress Rita Hayworth, and the artist Willem de Kooning all have in common?

They all fell off the cliff, near the end of life, into the abyss of Alzheimer's disease.

How did they arrive at the cliff's edge? What changes led them to that perilous position? We often think that we're doing "just fine" right up to the moment that we discover that we are at high risk for falling into that pit. But in fact, for most of us the risk of a disastrous end-of-life epoch has been slowly growing in our brains year by year, beginning at about the time that our brain reached its peak performance capabilities back in our 20s or 30s.

In many older people, the lack of appreciation of the extent to which basic faculties may be fading is rather astonishing. My colleagues and I have conducted extensive surveys in people over 60 to determine their level of performance when driving a car. Most of those surveyed fervently believe that their driving is just as good as ever. From their perspective, the problems they have while driving is that all those other people are not good drivers! In fact, objective testing shows that a substantial majority of drivers over 60 see and respond to driving hazards more slowly and with poorer clarity as compared with younger

66

drivers. They also have substantially higher risks of a serious automobile accident. Still, most remain oblivious to their growing impairments. In the same way, most older individuals underestimate the extent of their cognitive decline in other domains—and interestingly, the older they are, they greater their errors in self-calibration.

Scientists have documented age-related cognitive decline by recording changes in several hundred elementary behaviors and abilities, in large and varied populations, across the entire lifespan. Just about every aspect of our brain's operations—its recording, interpreting, identifying, deciding, and reacting—slows down. Just about every aspect of a brain's accuracy—precise responding, agility, fluency, and reliability—slowly deteriorates. More often than not, that gradual decline is accompanied by a loss of confidence, a slow retreat from full engagement with the rapidly changing world, and a slow social withdrawal back to a more egocentric personal life.

How rapid, and how substantial, is the decline? At roughly 55 to 60 years of age—30 years beyond the peak performance age—the average person has moved from the 50th percentile in cognitive performance to about the 15th percentile. At 80 years old, that same average person's cognitive performance has dropped to about the fifth percentile.

I shall later consider, in detail, how those changes relate to how a person uses their brain across these multiple decades of decline. I will explain how the changes in the chemical and physical brain that bring you closer and closer to the edge of catastrophe arise as a consequence of natural, plastic changes occurring in your own brain. Of course that bleak discussion is a preamble to explaining what you should be doing to sustain your cognitive abilities at a higher level—and about what you could be doing to maintain a safer distance away from the edge of near-end-of-life disaster.

One rather remarkable aspect of the changes recorded in human brains and in performance abilities is the surprisingly large variability in the patterns and magnitudes of decline. My wife and I have an old family friend who has a delightful 100-year-old dad. Harold still works eight hours every day as a hospital volunteer. He's a competent driver who serves as the local taxi for a large number of his (mostly much younger) pals. Harold is full of humor, full of insight, and full of information and understanding that can only come from a long history of good remembering. He's chock full of life. As with Avram, the nonagenarian inventor from chapter 2, I'm pretty sure that if I measured Harold's brain speed or fidelity or accuracy, he'd still be in the range of an effective late-middle-aged individual. Of course, a little less optimistically, we also know other individuals who have dramatically slowed down in their physical and mental lives at far younger ages.

While our brain's ability to accurately and rapidly interpret, record, and use information from our senses or memories to guide our actions is in decline in most mature individuals, other abilities are still growing into older age. If we test a large sample of citizens to assess their knowledge as a function of age, they score higher on the average, decade by decade, into their 60s or 70s. Knowledge is cumulative; you have to live for a while to build up a really large pile! Unfortunately, while it may still be growing, the acquisition of knowledge usually slows down in your older years because the brain's recording (remembering) abilities are slowly fading. I pointed out earlier that an average 25-year-old remembers the majority of things that they have just heard or seen. An average 65-year-old forgets more than they remember. Unfortunately, their personal growth is usually slowly decelerating.

An older brain has also spent a lot of time manipulating that very large body of information that it has earlier recorded in thought. That mental manipulation commonly results in a depth of understanding that is empowered, again, by a long history of direct experience and mental exercise. We call the product of this mental exercising "wisdom." Our human history is replete with older, wiser women and men who could not have led us or helped us to the same degree in their youth. Imagine a 25-year-old Winston Churchill leading a wartime British Empire, a 25-year-old Mahatma Gandhi transforming the Indian nation and Indian society, a 25-year-old Mother Teresa establishing one of our greatest examples of selfless human charity, or a 25-year-old Nelson Mandela leading the South African majority out of bondage, and you begin to get the picture. If your brain is well taken care of, then deep understanding and intelligent resolution can come from—and be absolutely dependent upon—your being around for a considerable span of time on Earth!

As I have earlier noted (and shall later discuss in more detail) this long progression of sliding back down the slope is a kind of neurological reversal for the brain, moving it backwards toward the operational characteristics that defined our physical, chemical, and functional brain during our infancy and early childhood. As the brain slowly regresses, the greatest losses first occur in the brain areas that were the last to come online. Because these areas contribute so strongly to our fast mental operations and our flexible intelligence, it is very important that we sustain them as long as possible. The competitive successes of most older individuals is the result of an interplay between their steady growth of knowledge and their decline in the accuracy and speed with which they can manipulate it. For some decades past the peak, knowledge can win out. Ultimately, for most of us, the deterioration of our brain's performance abilities will ultimately dominate.

Who wouldn't choose to continue to grow their resources of knowledge and ability to the end of their life, if it was within their power to do so? I hope that you are beginning to understand that almost all of us do have that power.

For further explanations and extensive references and citations related to the information in this chapter, please visit www.soft-wired.com/ref/ch12

13

BELIEVING IN BRAIN PLASTICITY

Applying this Science to Recover a Life

The accident occurred, as they so often do, at a moment when all seemed right with the world. Ryan had enjoyed dinner with his friends and his girlfriend Jackie, and was racing in his boat back across Lake Travis to help everyone get back to their car on time. He felt wonderful to be happy and secure with friends on this warm summer evening out on the lake, and knowing he had full life ahead of him.

Suddenly, out of the darkness, in the deep shadows, a large black boat with no lights loomed in the foreground and was suddenly right in front of Ryan, bearing down on him, tearing his world apart, in a flash.

Ryan was luckier than his friend's fiancée, whose body was never found. But although Ryan survived, it was with grievous injuries. It took a surgeon several hours to extract the many pieces of shattered bone from his brain. That procedure required the removal of a substantial part of the cerebral cortex and other brain tissue. Still more of his brain tissue had to be removed a few days later to reduce uncontrolled swelling. The suggestion that perhaps Ryan should be "let go" hung in the air for Ryan's mom and dad, as the neurosurgeon explained that even if he recovered consciousness and survived, he was not expected to recover speech understanding or to ever speak again, and would probably not be able to take care of himself in even the basic ways.

Ryan's parents chose life.

When he recovered his consciousness—and for many weeks afterward—Ryan could not talk. He could initially respond in only very primitive ways. He could not control his movements. While Ryan could remember things from his past life—like his great affection for his parents—he struggled to record new memories. As in the movie Groundhog Day, every new day had to be restarted, as if all of those earlier days had not been lived. The prospects for recovery seemed bleak.

Ryan's brother, mother, father, and girlfriend spent much of every day with him, coaxing and coaching him in an attempt to initiate the recovery of his movements and his speech. Although he could not remember, Ryan did begin to show that he at least sometimes understood what people were asking him to do. After weeks of effort, it became clear that he was trying to mouth the sounds of words, although it was many weeks more before he could actually vocalize the sounds himself. Through months of effort, tiny step by tiny step, he gradually recovered a limited but useful functional vocabulary. Following the same torturous path, he also slowly re-established the ability to control his basic posture and movements.

As Ryan's family members and therapists worked hard to help him achieve these small daily victories, his dad left his job as a successful construction company executive to dedicate himself to finding therapeutic approaches that might contribute to his son's functional improvement. After following the branches of several paths to try to help Ryan, he contacted me to ask whether or not the neuroplasticity-based training programs that my colleagues and I had developed could be applied to help his son. After reviewing Ryan's records, I encouraged their use.

Ryan and his family almost immediately saw major gains attributable to the use of these intensive computer-delivered, brain plasticity-based training programs. Over a period of about two months, he substantially recovered his ability to remember and to operate with a reasonable level of cognitive control. Over that same period, Ryan took a big leap forward in his language and attention control abilities. One consequence was a rapid re-establishment of his ability to operate more independently in the world—because now, this young man could leave his home on an errand, remember its purpose, achieve his goals on the errand, and remember how to find his way back!

I recently had a conversation with Ryan on the phone. He's doing very well. He has no difficulty in following a conversation. He answered my questions and injected a little humor in his responses. Ryan is now working with his dad, and living in his own apartment with Jackie—who is now his fiancée. He enjoys life, and once again, has a full, rich measure of it—to live independently, but with the people that he loves around him—with any luck at all, far out into a recovered future. This outcome is all the more incredible because so much of Ryan's brain mass is missing that he literally has a hole in his head.

Ryan and his parents have seen brain plasticity in action. We know that Ryan's recovery has occurred—and could only occur—through a massive level of connectional remodeling in his brain. We know that his brain has found new ways to achieve very important things that were formerly only achieved in brain areas that Ryan has lost to injury.

The remarkable capacity for brain change, achieved by the application of appropriate, intensive training accomplished through the patient guidance of his parents and skilled therapists and restored through the use of an intensive software-based brain training program literally gave Ryan his life back.

Ryan and his parents believe in the power of brain plasticity because they've seen it in action. Perhaps you, too, are catching the wave.

For further explanations and extensive references and citations related to the information in this chapter, please visit www.soft-wired.com/ref/ch13

PART THREE:
CREATING "YOU"

14

YOU ARE SPECIAL

How Your Brain Remodeled Itself to Create the Unique Person You Are

In reflecting on our losses as we age, our natural tendency is to think of changes within ourselves in terms of our superficial performance abilities: How fast can I run? How much do my legs ache? Why can't I remember that word? Why can't I remember which way that goes? These changing behaviors also reveal something else that you may come to regard as even more important that is changing, which is yourself—your essence, your soul, your "you-ness."

Perhaps, like me, you have witnessed the slow decline of a beloved parent in the grips of old age and senile dementia. You may have watched this person progressively decline into a sort of "second childhood," during which they lose their connection with the world, simplify their history to a distilled set of the best old memories, and shed innumerable things that once defined them as the rich, unique characters that they were. An important objective of a well-ordered mature life is not just to sustain, but to continue to enrich and grow the person that dwells inside your skull and body. To understand how to achieve that, it is useful to understand how that person has grown within you, as a product of brain plasticity, within your lifetime.

As I described earlier, when a baby is born, there is really not much evidence that a unique person is in residence; that person is created by brain plasticity. How is he or she established? As you acquire each skill and load up your brain with the representations and meanings of the things of the world, everything in every moment that the brain has learned or remembered is recorded in the context in which it occurs. On that basis, as we have earlier described, the brain physically constructs what goes with what in our world. It builds a massive encyclopedia of information about those innumerable categories of things that naturally belong together, and that thereby predict the occurrence of one another (their "remembered associations"). Given that information, the brain is making continuous "predictions" about who or what belongs in the scene, and about the set of things that could likely happen next. This forward prediction is the motor that propels your "stream of consciousness."

For example, if I hear the first two notes of "Happy Birthday," I am beginning to predict the third. With three notes, I have a still stronger prediction of the fourth. With four notes, most listeners are led to remembering the fifth, and so forth. Seeing an image of a cow leads to a prediction of things most commonly seen or heard with cows (e.g., other cows, pastures, mooing, fences, cud-chewing, barns, rural vistas, hay). In the background, but always ready to arise to consciousness, there are hundreds or thousands of less frequently occurring associations that also belong with cows but are not in that most common category (e.g., cream, a milking machine, a picture of a cow with cuts of meat represented by dotted lines on its flanks, a barnyard cat waiting for milking). If I have remembered this information under emotionally charged circumstances, associations are more vivid. Fear of cows might lead my brain directly to their horns, hooves, or biting maws. Sympathy for cattle might lead me to think of their big eyes, their gentle nature, or the horror of their slaughter. Our construct of the world is fabricated from hundreds of thousands or millions of these learned associations. If we record from the brain while they are being established, we actually witness these associations being constructed in the physical brain, through brain plasticity.

You can think of your brain and memory as operating somewhat like a modern Internet search engine, where the top-ranking search outcomes (your easiest-to-recall memories) are identified by the frequency and the importance of their associations with the research entry that you have just typed in (noticed or thought about).

Something else very wonderful is achieved as a product of your brain's powerful information-recording/retrieval system. For every touch, for every moment of vision, for every sound or word heard, for every moment of thought or action, there is a second very reliable concurrent association. That second association is to you. You are always in the picture, as the associated source of all sensations, thoughts, and actions. Through literally billions of brief moments of brain plasticity, the predicted, concurrent "you" grows stronger and stronger and more and more predictably and reliably referenced.

Because billions of touches strike your body's surfaces, that "self" that grows within your brain is embodied; you view yourself as co-extensive with the limits of your skin. If you ponder about where you are, as a visual witness of your world, you place yourself in your head directly behind your eyes.

The person that grows within our brains is inherently selfish. We are primarily self-referenced receivers of, users of, and actors on the specific things that we sense and learn about and think about and act upon, and the specific skills that we achieve, through the epoch of our lifetimes, in our limited culture, in this era of human history, in our place in the world. However, when we have external sources of self-reference that are particularly frequent or that carry special emotional weight, we freely extend our personhood to incorporate other essences into our own. A husband or wife and the children that they nurture can thereby become an inseparable part of the person created within their brains. You really do come to carry a brain-constructed attachment to your child. And you really do come to (at least spiritually) resemble your mate or your beloved shih tzu! The familiar surroundings of a home, work site, church, tavern, political affiliation—or any place else where you spend a lot of time, or any other thing you grow really fond of—can also literally physically grow, through association, into that special person that is you.

Of course you are also defined as a person who is good at many specific things, and (of course) inexperienced or not very good at a million others. You have experienced and recorded millions of events and missed out on trillions of others and you can recognize and name the many thousands of things that you have experienced in your world on both a "real" and symbolic level. You have richly associated those things of the world with one another, to achieve an understanding of what goes with what within it, and you have manipulated that information in millions of additional ways to generate ideas, inventions, and predictions.

Scientists and philosophers have spent a great deal of time trying to understand how to account for the operational elaboration of the person that you are. The genetic determinism that seems to be so strongly buttressed in cultural anthropology and biology in the arguments of modern synthesizers like E.O. Wilson, Richard Dawkins, or Stephen Pinker argues that humans are endowed with the essential capabilities that define them as human (as opposed to bovine or feline or ursine or ovine), and that our evolving genetics underlie our species' culturally evolving behavioral strategies and expressions. In this view, humans talk even though the apes did not, because we have the genes that enable talking and understanding speech. Humans have a capacity for higher mathematics while the apes are stuck at the level of counting fingers and toes because we have genes that enable it.

Their view is correct to a point. Homo sapiens have doubtlessly genetically evolved, for example, to favor traits that are better suited for settled agrarian societies versus roaming hunter-gatherer societies. On a basic emotional or social level, behaviors across different human societies have more commonalities than differences, and there are fundaments of language structure, repertoires of motor control, and overlapping aspects of emergent cultures that certainly do reflect the inherited endowments of our bodies and brains. Our ability to produce speech is shaped by our inherited control of the jaw, tongue, lips, pharynx, larynx, and chest wall. The possible ways in which we can dynamically shape produced sounds are limited by and determined by these inherited resources. Lucky us, that our Adam's apple (larynx) has dropped lower in the throat, and that we can voluntarily control our breath, so that we can alter sounds more usefully than any other primate by controlling the airway above it! It's a small difference from the chimp, but for speech production (and with our larger, highly plastic primate brains) it made all the difference. That difference ranks right up there in importance with our adoption of an upright posture, which had a dramatic impact on hand (tool) use, once we weren't walking on those forepaws all day long!

Many inherited features of our brain are also in play, as determinants of how sounds and sound relationships will be analyzed by our brains, about how emotions will flow from our reactions to social circumstance, on the tastes and smells that guide our feeding, among many other relatively predictable behavioral characteristics that define and distinguish our species.

At the same time, what a pathetically limited view of humankind and of ourselves we come to if we think of ourselves as rigidly determined by genetics—when arguably the most remarkable thing about us is the incredible way that our brains and environments generate individual differences and behavioral exceptionality amongst us.

Vive la difference!

No two of us are exactly alike in our unique abilities and information resources that each of us acquires within our lifetimes. Each of us has done a one-of-a-kind job in loading our brains with a personal repertoire of skills, and with our own private and absolutely unique library of information. The notion that humans had to mutate each time we made a significant cultural advance once again strongly underestimates our greatest inherited resource—our unprecedented capacity for brain plasticity.

After all, if it were really the case that genes were above all else, then why doesn't the average citizen of Manhattan, Tokyo, or Berlin fit in better in a Yoruba society? Why does learning to ride a horse actually have to involve the use of horses? Why does getting a physical beating every day give you an enduringly bad attitude, even while your genes were inclined to give you a naturally sunny disposition? Why couldn't those Greek geniuses make more intelligent use of calculus? Was it because they hadn't yet mutated in the crucial ways that enabled mathematical proficiencies? And what genetic changes could possibly account for the remarkable cultural changes expressed in our world over the past two or three hundred years? Our individual development stems from the specific uses of our inherited brain resources. Without those uses—without all those million-and-one brain-changing experiences— we'd have nothing.

Let's consider a simple real-life example of the innumerable ways that we commonly confuse the contributions of nature (our genes) with nurture (the specialization of our brains derived from our individual experiences). Little Bonnie shows up at school, already reading a bit, eager to read more. She's a bright little penny, engaged by almost everything about learning. She loves school. Little Buddy, on the other hand, is quiet and distracted. He doesn't show much interest in learning letters or phonics, and it is clear that reading will be a struggle. He almost immediately understands that he is not going to be good at school.

We now know that on the average, a child like Bonnie has probably been raised in a supportive and language-rich family. By her first day of school, Bonnie has most likely heard six to ten times as many words, expressed in far more complex language forms, than has Buddy.[3] Even more importantly, Bonnie's generally thriving family probably supported

---

[3] Note that there are many children with genetic faults whose slow development cannot be attributable to variations in early childhood experiences. Still, for the majority of struggling children, experiential impoverishment—or, all too often, a history of neglect or abuse—has contributed greatly to what is later expressed as a child's all-too-predictable academic failure.

and affirmed early language and early exposure to books and other teaching tools. Dozens of times a day, Bonnie's parents probably expressed some form of "Atta girl, Bonnie!"

In contrast to Bonnie's generally thriving home, Buddy's home life could be described as generally struggling. A large proportion of the sparse conversation that occurs in a language-impoverished home like Buddy's is negative and corrective. Research has shown that year by year throughout his early childhood, Buddy's conversations with his parents have likely more often than not resulted in ridicule and negative feedback. This kind of interaction occurs, on the statistical average, when families don't talk very much, have lives full of stress, and aren't doing well overall. Buddy's language-sparse kin are much more likely to say "Knock it off, Buddy!" than they are to encourage or support what he has to say. When Buddy starts school, it feels to him like everyone is incessantly talking there. He has learned to associate language with trouble. He's learned to avoid it while Bonnie has learned to love it.

Reading is the exposition of language in its most elegant form. Bonnie has been advantaged in her use of language, and in all the benefits for scholarship that come from it. Her rich exposure to language through reading further extends and amplifies her neurological gains. Before long, she can be expected to read several million words a year just for fun. Unless he's really fortunate, Buddy can be expected to fall further and further behind. Studies show that if Buddy is an American kid with a typical language-impoverished history, he will be five or six years behind Bonnie as a reader by the time that he enters high school. At this age, Buddy will be doing almost no reading for fun. And by that time, Buddy won't have a very high opinion of school, and insofar as his opinion relates to his academic performance, of himself.

We all instinctively cherish a young child like Bonnie who has grasped life so firmly by the tail. But how should we think about Buddy? His inherited resources may or may not be as strong as his successful classmate's. At the same time, we should never forget that the variation in the quality of human experiences and learning, down to the individual local environment of the home and family is a second—and usually even more crucial—developmental endowment. They are not genes, yet still assuredly passed on without the child having anything to say about it. Differences in the quality of "nurture" regularly make profound differences in physical and functional brains, and in the failures or successes of the people who own them.

It was partly a concern over the fates of children like Buddy that led me and my colleagues to develop intensive therapeutic training tools (called Fast ForWord) to help kids like Buddy. There are countless kids with language and learning impairments, and our research found that the right kinds of brain training can help them develop more normal language and other mental abilities. With that jumpstart, they can take off as successful readers.

Reading success can rescue a child like Buddy. Several million children with impaired language and reading have used our computerized, brain plasticity-based, game-like training programs and they have made substantial gains in academic achievement. Most of these children become more accurate listeners. Speech reception and usage becomes more sophisticated. They talk more. Their memory improves. Their thinking sharpens. They read faster, and more competently. Their confidence receives a major boost. Their brains advance and become better-specialized at receiving language, at reading, and in indices that measure their intelligence. Most of all, their prospects for a productive and rewarding life are significantly elevated.

Therein, once again, lies the great hope of brain plasticity. Children who have a brain that has been disadvantaged by a limited experiential history or by genetic weakness aren't stuck on an inalterable path to failure. Whatever the circumstances of a child's early life, and whatever the history and current state of that child, every human has the built-in power to improve, to change for the better, to significantly restore and often to recover. Tomorrow, that person you see in the mirror can be a stronger, more capable, livelier, more powerfully centered, and still-growing person.

For further explanations and extensive references and citations
related to the information in this chapter, please visit
www.soft-wired.com/ref/ch14

15

YOUR PARALLEL UNIVERSES

Growing, Sustaining, and Controlling All of Your Brain Worlds

I've mostly talked about your brain adapting, plastically, to create a model of that Great, Big, Wonderful World that we live in, and about how your brain controls its own plasticity to gradually figure out how to operate successfully within it. It's time to deliver a little more shocking news to you. First, that world in your brain isn't exactly real. There is no real "orange" in a sunset, "sweetness" in that cookie, "amusement" in watching that kitten, or "wonder" in contemplating your navel or the cosmos! Your world is a brain world. It made it up. Fortunately for you, your brain calibrated its construct of the "real world" so that this grandest of fictions makes sense. Sort of.

You may already have a sense that the world in your brain isn't strictly real, but you might be less aware that your brain has created more than one world. Several other universes beyond the one "out there" have also been constructed inside your skull. Because an understanding of these parallel brain worlds is key to your brain health, I'd like to tell you more about their natures.

One of those worlds that it is really important for you to master is a complex neurological construction representing your body. That world extends from the tip of your toes all the way up to the top of your head. I'll call it your "corporeal world."

About 30 years ago, I had been increasingly ill over a period of about a week, but had been trying my best to ignore it. I was feeling poorly and had a high temperature, and had already decided to drop in at a university clinic near my home the next morning. I'd worked late into the evening that Friday, then come home and just sat on a chair in the corner of my kitchen reading, drinking little sips of milk to try to soothe my raw throat. At some point near midnight I noticed that I could no longer feel the milk in my throat. Within perhaps ten minutes, I slowly realized that all of my achiness and throat pain had evaporated into thin air. Very soon afterward, although I was still awake and aware, I had absolutely no feeling or sense of my physical body.

Up to this point in my life, I had not thought much about the fact that I had a mapped reconstruction of my physical body ever-present in my brain. My guess is that you don't think very much about that either. When it departs suddenly, you immediately know that a very important actor has left the stage! In a mild panic (and jumping to the conclusion that I had just had an aneurysm or stroke) it took me a minute or two to calm down, remembering that I could hardly lose the feelings from all of my body from a vascular accident and still be alive. Still, I nervously called up to my wife Diane who was sleeping upstairs in our bedroom to let her know that something important was happening, that she should dress because I would need her to help me get to the emergency room two blocks away.

Can you imagine walking those two blocks with absolutely no feeling coming from your body? There was no physical me there. I floated across the landscape as a disembodied agent, a conscious presence located just behind my eyes about five feet above the ground, gliding step by step as a kind of insubstantial essence. I still vividly remember that fascinating ethereal passage of my disembodied spirit on this short trip through the cold night to the ER.

It did not take long for the ER staff to diagnose severe pneumonia and put me on oxygen. As a part of their examination, the nurse drew blood from my arm. I could feel none of her jerking my arm into position for the blood draw, and I did not feel the needle until out of nowhere came a pulse of deep pain. "You have a classic case of pneumonia," I was told. "Your blood oxygen level is dangerously low. You're going to have to spend several days in the hospital."

About a hundred years ago, a British neurologist showed that you could block all feelings from the skin without eliminating sensations of subcutaneous (deeper) pain by decreasing the skin's supply of oxygenated blood. When you have an illness like pneumonia, the body knows that it has to deliver more blood to the parts that are most critical to keep you alive, like your brain, heart, lungs, kidneys, and liver. It steals blood from your body surfaces, shipping it to that vital core. In my pneumonia, there was so little oxygen delivered in that reduced blood flow to the sensory tissues in my skin that they had just shut down. I had lost the embodied, corporeal Mike!

I lay in my hospital bed for about two days before the antibiotics killed enough of the bugs to allow the oxygen levels to rise. At that point the superficial tissues of my body were re-oxygenated enough to restore body sensation. Then, over a period of more than a day, different regions of my body would again show up, then fade away. "My right leg is here again," I would smile. "Oops. It's gone again," I'd frown. All in all, it was an eerie and unforgettable experience.

As a side note, I might add that when a medical school professor shows up in the university's hospital with a classic form of any disease, the attending physicians take great advantage. Every young doctor in training has to understand what a classic case of pneumonia is all about. A steady line of earnest young men and women trooped through my room, listening and thumping, with their cold stethoscopes placed here, then there, on my back and chest. I could hear their sighs and assents of understanding, as they heard and felt what it was like to have one's raspy lungs largely filled with fluids.

There is a lot more to your brain creating reconstructed maps of different things happening in your body than your skin sense. Your corporeal world is a lot more complicated. Still, touch and its companion, your sense of temperature, have special status, because they rise so powerfully to the level of consciousness, and because you employ your hand and your mouth in such wondrous ways as exploratory instruments. In this respect, humans have been greatly advantaged by adopting our upright postures so that our hands are free for tool use, making us all the more adept at bringing things to our mouth for further inspection. But underneath those wonderful cutaneous senses are magnificent contributions to your corporeal self coming from an equally impressive array of "sensors" embedded in your deeper tissues, and in your muscles and joints. While you don't directly feel the sensors dedicated to controlling your movements, if I had lost them in my pneumonia they would have had to carry me to the ER on a stretcher! The brain is smart enough to keep the complex information coming from your muscles and joints *sub rosa* because the million-and-one small adjustments in muscle fiber activation that they control are too complicated to make sense to a perceiving and conscious person; you have a lot of muscles, tendons, fascia and skin senses contributing to the control of those multiple dimensions of movement around a lot of joints. Instead, your corporeal world creates a beautiful, ongoing sense of your body position in 3-D space that's known as "kinesthesia." These otherwise-unperceived sensory sources bombard your spinal cord—and through it, your brain—with more than a million lines of information that continuously add critical influences to the construction and actions of your corporeal self.

You also have lots of other silent messengers coming into your brain from every organ—in fact, from every crack and crevice in your physical body. Those organ senses are, again, mostly unfelt. Pain input is the universal exception. Everywhere you can think of, your body has been stitched with incredibly delicate little wires that are just waiting, in a sense, for something bad to occur. In mild forms, they confer achiness and the small distresses of neck, body, hip, shoulder, and limb. With more damage, these little devils can gang up and drive you to agony. Except in the skin itself, this diffuse network of tiny sensors doesn't provide information that the brain can locate precisely. Your pain sense "borrows" from the skin senses to alert you about your painful insides, so that you and your doctor know roughly where the trouble is! In most older individuals, these ongoing internal irritations are continuously with you, as a myriad of vaguely located aches and pains that can arise anywhere between the top of the head down to the tips of the toes.

Another universal presence is the wiring that controls the flow of blood to different body tissues. Some of these wires are shipping back information from vascular system sensors. Many more are providing fine-grained control of your heart, and of your arteries and veins, almost down to the level of your capillaries. In the body as in the brain, when a particular organ (skin, muscle, kidney, liver, or naughty bits) is active, there are mechanisms that assure that they get more blood to help them in their tasks. When you're using your physical body, blood flows away from your digestive core (which can almost always wait awhile to deal with your last meal) out to your muscles and skin. When you quiet down and rest a little, blood flows back toward your innards, and your skin and limbs "cool down" as they come off line. Your brain is making very important contributions to this crucial, ongoing control. How crucial? Organ death (and its owner's probable demise) is signaled by its loss.

As you're reading this, I want you to stop for a minute and focus on the feelings that are coming from your body. Shift your mind to focus on the feeling coming from the index finger on your left hand. Come on, take a minute and really focus. Now think about the knuckle on the back of your hand at the base of that index finger. Can you feel it? Now close your eyes and imagine the outline of your hand, in your mind. If I asked you to draw it with your eyes closed you could do it, and within the limits of your drawing skill, your depiction would be accurate and just the same size that your hand is, in reality. Now move the focus of your feeling to the ribs on the left side of your body; to your right knee; to the heel of your right foot; to the left side of the tip of your tongue. Were you already aware of the fact that you are being continually bombarded by sensory information from your corporeal world that, as long as you're awake, and unless you are in a very deep, practiced meditative state, you can never entirely shut off? If you don't believe me as to this last point, try telling yourself: "Okay, RIGHT NOW, shut it off."

You can't, can you?

When we think of our brains and bodies, our first thoughts usually go to the control of our movements. Actually, we're much more complex and interesting than that. While the brain warms up to your laughter, most of laughter's feelings are coming from the brain's reconstruction of its many expressions coming back from our chortling, convulsing bodies. We shudder, smile, shiver, quaver, wince, itch, giggle, tickle, sigh, frown. We feel the dull ache of a full bladder, the vibrations of the passing trolley, the hitch in our knee, the contraction in our chest, the crick in our neck. We feel the worry and the relief, the kindness and the cruelty, the problem or the solution coming from our corporeal world, just as we feel it in that more ethereal world of our thoughts.

Very few of us understand the extent to which our brain plays a key role in regulating body functions. Some of this influence is achieved by its release of chemicals—regulatory hormones—that, like the release of chemicals in the brain itself, stimulate important body processes. Other influences, also in play for every organ in the body, are electrical, and direct. You may have forgotten that you actually learned to urinate and defecate in the proper way (i.e. only at certain times and only in certain places). You might not have realized that you're eating too much because the brain, on the basis of information that it receives from the body, is telling the corporeal you that it really likes pie!

When we discover that something is failing within our bodies, our first thought is that those failing parts or processes must be repaired in situ. "I have a weak bladder," Ann might tell herself, with no blame given to the brain in charge. "My knee is stiff," Conrad may hear himself saying. "Maybe a little more exercise will loosen it up again." Sure, that might be a good thing to do, but had it occurred to him to ask the brain to help? Why has the brain shut down the mechanism responsible for ample delivery of the blood to keep that knee joint lubricated?

Of course if you're operating with full control, your corporeal self does a good job of communicating with your other worlds. When you read the word "pinch," can you feel the quiver in the thumb and finger of your dominant hand? Or can you feel a slight sensation of being pinched? Most people do. When you're watching a movie, every time there is a hand or a foot or an ear in a scene, the parts of the brain related to the representation of just those parts of your corporeal self light up like a Christmas tree! That's because there are close linkages between the brain's reconstruction of our outside world and its equally special representation of our body in our corporeal world. All and all, for your health and survival, keeping the status of your corporeal world in good stead is just about as important as is keeping your reconstructed understanding of that external "real world" under control. I will talk a lot more about how you can achieve this later.

Just as your body sends ever-present information about it to you, so, too, does that external world. Our external world is not just about seeing. We swim in a sea of air in that world. We have an incredible ability to detect and—through our brains, identify—volatile chemicals in our ocean, through our sense of smell. The loss of this ever-present ability strongly impacts the pleasure of eating, and our ability to detect many things that we would otherwise avoid or be attracted to in our external world. If all is going well in our brain, our external world is odiferous!

Moreover, anything that moves in that sea of air generates vibrations that continuously bombard us. We detect those vibrations, again, with incredible sensitivity; our external world is a world of ever-present sound. If you don't believe me, stop, right now, and listen.

In my earlier research life, as I've related earlier, my colleagues and I worked hard to help people who had lost the ability to detect air vibrations—their sense of hearing. Scientists have argued that the loss of hearing is even more debilitating then the loss of sight in humans, because hearing loss can so greatly attenuate our social communication abilities. While humans can thrive if they develop effective non-auditory means of communication—as do the congenitally deaf, with sign language—most individuals who acquire a hearing loss after a period of normal hearing are devastated by it because it results in their social isolation. It's not just about communication. Life is significantly impoverished if we suddenly—or, more commonly, slowly—lose access to all of the hidden information provided by hearing and smelling our world.[4]

---

[4] There are many among us who have been denied the rich contributions of hearing by genetic or accidental circumstances in their lives who have found a Visual World strategy to overcome those losses. The ideal human should work to understand at least a little of the visual poetry of their sign languages, just as we work to understand our and other sound-based languages.

A guy named Sid was one of the first people in recorded history to gain insight into a third brain world. He did this by trying his very best to exclude the influences of our brain plasticity-created constructs of the external (i.e., visual, olfactory, and hearing) world and of the corporeal world in an attempt to understand what was really inside his skull at the core. Through this process, Siddhartha Gautama found very interesting things there, more than 2,500 years ago. The great Buddha's followers (which in different traditions later included many Christians, Muslims, and people of other religions) almost immediately recognized that they could change the nature and the power of these core qualities and abilities by forms of inwardly focused mental exercise. This third world is a world of purely mental essences, of qualities that we use to color the otherwise emotionless events of our life. Things would be pretty drab for us humans if there were no light in our lamp, and if we could not love or care, or know or puzzle. You may be thinking, "Sure, but doesn't every human have those abilities?" Indeed, they do, but in most cases they are not terribly refined. How bright are the ways that you color your life? How many hues are loaded onto your palette? How richly have these core abilities been advanced by your brain plasticity?"

Of course there are many other possible worlds that you may have created in your brain. Those worlds live in the realm of the fantastic, which is another way of saying you (or someone else) simply made them up. There might be fancy names for some of those worlds, like "higher mathematics," or "philosophy," or "neuroscience," or "investment banking," ot "World of Warcraft," or "FarmVille." There are thousands of these landscapes that can be constructed in your mind, loaded with mountains of abstraction, mythic creations, logical elegance, or possibility or impossibility. Once they enter these worlds, there are many individuals who spend much or even most of their time in them. "Out to lunch," or "I'll be back later," is the sign that many of these individuals might put up on their mental door.

You are the integrated sum of these worlds. You live in all of them. By your historical experiences, you refined all of them. You are them.

For further explanations and extensive references and citations
related to the information in this chapter, please visit
www.soft-wired.com/ref/ch15

16

LOADING YOUR OWN ENCYCLOPEDIA MAGNIFICA

Memory Acquisition and Use

Abraham Lincoln apparently never forgot an amusing anecdote or story. He seemed to have thousands of them held in memory, and was famous for retrieving just the right story to fit the circumstances at hand, with almost no repetition. According to the several thousand people who thought that Abe was one of their best friends, he was a master at getting that punch line just right! Lincoln's remarkable capacity to dominate the men around him through perspicacity and great good humor stemmed from a truly remarkable memory.

Napoleon Bonaparte had an equally prodigious memory that he applied to significantly different purposes. When he heard that Admiral Pierre-Charles Villeneuve had fled from the English fleet just as he was supposed to be organizing the protection of Napoleon's army's invasion of England, the Little Corporal flew into a fit of rage. He made an immediate decision to do an about-face with his army (scattered over more than 500 miles up and down the French coast facing England), turning eastward to invade the Austro-Hungarian Empire. Yesterday, all of his extraordinary mental energies were focused on conquering those damnable British. This morning, he learned that he could not do that. This afternoon, out of his very powerful brain came the detailed plans for invading the territories of a second enemy.

Over a several hour period, hardly pausing to catch his breath, Napoleon dictated the order of march for all of the units and of all of the necessary arms and supplies of his scattered army in complete detail, and identified the sites and dispositions of the expected battles—which all came to pass. Without a hitch and without any significant error in planning, 200,000 men marched east toward a predicted series of victories culminating, just as he planned, on the fields of Austerlitz.

Napoleon's genius was clearly dependent upon a prodigious memory.

Kim Peek had a much better memory than either Emperor Napoleon or President Lincoln. The astounding Mr. Peek could remember every chapter and every page and, in fact, could accurately reiterate every sentence printed in a thousand books. Later memorialized in the movie "Rain Man," his legendary memory did not translate into great personal achievement. Memory is a key product of our powerful self-organizing brains. At the same time, there is more to a successful life than a good memory. Remembering is useful only to the extent that it evolves in ways that confer understanding, and can be translated into useful thought and action.

You have recorded and stored a massive amount of information in your brain. For starters, you know the identities of—and the words that symbolically represent—many thousands of things in the world, as well as their qualities, possible actions, and relationships with one another. While it's true that you may sometimes have difficulty coming up with just the right name or word, just think of all of those names and words that you can retrieve when you have to! You have a pretty remarkable reference book at your disposal. You learned all of those words and their "meanings" through hard practice in life, as they were etched in your brain through the processes of brain plasticity.

It might be just a little discouraging for some of you to know that a typical 30-year-old has command of the use of about 30,000 words while the typical 80-year-old uses only about 10,000. Associations between words and their meanings are multiples of these numbers; the associative abilities of the typical 80-year-old are very impoverished relative to a 30-year-old. This is another good reason to avoid being merely "typical" at age 80!

Regardless of whether or not you are operating with a full memory deck, you can also relate your memories to one another in hundreds of thousands or millions of ways. A simple way to demonstrate this remarkable ability of the brain is by thinking about any common thing or word. Let's use the word "car" (or the visualization of an automobile in your mind) as an example. What does the word (or the visual memory of) "car" bring to mind? You might start by trying to remember:

- all the different cars you and your family have owned
- all the complex parts of a car (tires, carburetors, ashtrays, tailpipes, brakes, gear shifts ...)
- the many different kinds of car (race cars, convertibles, sedans, taxis, toy cars ...)

- any number of things you might find in or near or under (or not in or near or under) a car (a driveway, a snake, your friends, an old soda can ...)
- the many, many places you've been in a car (your prom, the corner store, Florida, your Aunt Bertha's farm ...)
- the fun, funny, exciting and perhaps sad things you've done in a car (saw a bear, went on a first date, drove to Uncle Charlie's funeral ...)
- the endless variety of places where you might find a car (on a road, in a parking lot, on a movie screen ...)
- and almost countless other things.

Your brain has structurally created all of these remembered relationships through learning—that is, through brain plasticity. I might add that these powerful products of your plasticity-driven reconstructions of "what goes with what" are exactly what the aforementioned Kim Peek was not good at, and that weakness explained why all his memory of the words in all those books mostly led him to the next word. Unlike Mr. Peek, our memories empower our intelligences, through plasticity-built associations, to lead a more fully empowered brain in a rich variety of other interesting and problem-solving directions.

As you've thought about the things related to cars, it may have occurred to you that some of your memories about them are clearer than others. Some even very complicated things in your memory have been especially faithfully recorded in your brain. If I asked you, "Where were you, or when did you first hear, that first one, then a second airplane struck the two towers of the World Trade Center?" you'd probably be able to provide me with an answer. If I asked you (and if you're old enough), "Where were you when you learned that President Kennedy had been assassinated?" or "How well can you see, in your mind's eye, those events in the delivery room surrounding the birth of your first child?" or "Can you visualize the setting and ceremony of your wedding?" or "How did you celebrate your last birthday?" I'd also expect to hear a pretty good accounting. In each of these cases, it is almost a certainty that you remember some aspects of such events in an elaborate, mentally reconstructed serial context. Events flow. Things move in time, in your memory.

Interestingly, as you bring up strong memories, your replay machinery advances your mental reconstructions in little snippets, and things go by at a pace that roughly matches the tempo at which they originally occurred. A musical conductor can mentally rehearse a long piece of music that she has mastered, all in her head. If we ask her to mark the beginning of key musical passages, we would discover that they occur almost exactly in time just as they would occur in time if she were actually conducting her orchestra.

Of course we vividly remember great disasters and triumphs in our lives because they occur under highly charged emotional circumstances. The neurotransmitters that accompany those strong positive and negative emotions are the self-same modulators of brain plasticity. When they're switched "on," as in these greatest triumphs or disasters in our lives, plasticity is powerful, and we almost never forget. Our memories are thereby strongly nuanced by our brain's own determination about just how important every event is to us.

Several other factors beyond emotional context also contribute to the clarity of your memories. First comes repetition and its consequence, familiarity. The more you hear about or are exposed to something, and the more things that you hear or see that it is related to, the more likely you are to remember it. Your general state of alertness and attention are also very important contributors to how well you remember something. When I'm bright and on the ball, and especially when I focus on trying to remember something, I am better able to remember it. But when I'm half asleep or distracted or my spirits are low, I don't remember very well.

The facility with which I remember something is also influenced by the strength and clarity of the brain's activities that are evoked as I attempt to "record" it. If I degrade that strength and clarity of the brain's coded representation of what I am trying to remember—for example, by simply turning up the background noise or dimming the lights—memory is the poorer for it. Finally, the associational powers of the things that go with a memory increase the reliability of its recording. One of the great tricks for remembering is to associate what you're trying to remember with other things that you are not likely to forget, like numbers, or friends or relatives, or any other such hard-to-forget "hook."

As I will discuss later, loss of memory in an older brain is contributed to by losses in all of these normally memory-vivifying assets. Some recovery in remembering power comes from finding ways to strengthen these key influencers of memory recording.

Individuals with localized brain damage may have no specific problems remembering things in the moment (short-term memory) despite losing virtually all historic (long-term) memories. Our brain temporarily registers the steady stream of things that occur through each day, but does not usually record them in a vividly retrievable form unless conditions are right, and unless we make a conscious effort to record them. We are limited in how many things we can keep straight in our short-term memory (also known as "working memory") as the information scrolls by. If we don't act on these ephemeral memories within a minute or less, they usually fade away. Over that time, as we have related earlier, the brain activities representing working memory are fed back across our great brain systems when the brain is in a learning mode, to bias plasticity at every level.

Short-term recording is a prerequisite for long-term memory. Scientists distinguish our ability to remember and describe earlier-recorded information over the long term as explicit, declarative, or episodic memory. These long-term memories always have a forward energy because memories lead to other memories. These memory progressions are the product of associative memory.

Our ability to remember how to perform basic skills that we have learned earlier in life is often called "implicit memory" or "non-declarative memory," because these enduringly "remembered" abilities do not ordinarily rise to the level of consciousness. You remember how to do a million and one things that earlier you learned how to do on automatic pilot, such as using a fork, or tying a shoe. In the late 19th century, psychologist and philosopher William James pointed out the great value of building up a powerful platform of skills and abilities that operate without much attention or thought, because these implicitly remembered abilities provide the critical foundation for all higher mental and action-control operations. The broader and deeper this foundation of implicit or automatic abilities, the more magnificent the superstructure that can be erected atop it.

I have largely discussed the operations of the brain to this point with a primary consideration of how the brain recognizes and remembers "What's that?" or "What's happening?" or "What comes next?" but of course one of the most marvelous things about this wonderful machine is its ability to generate and control remarkably flexible voluntary actions of thought and movement. To demonstrate some of the principles by which the brain operates in this mental domain, imagine a pencil. See the pencil—this nice, sharpened yellow pencil—in your mind. Grasp it in your hand (in your mind). Now write your first name in cursive with the pencil, in your mind. Can you do that, with your mind?

Now imagine a map of North America. Can you see it, in your mind? Place your imagined pencil tip down on Los Angeles. Now draw a line on the map, as you move it (on the map in your mind) to Montreal. Did you go through Chicago or Detroit? Brrrrr! Let's warm up, by moving the pencil, on the map in your mind, to the Florida Keys. Thank goodness it's warmer down there!

You might note that you certainly would not have required my instructions to go through this kind of mental exercise in your own mind, because you manipulate information like this in thought all the time. Moreover, you know that thoughts like these can be remembered and retrieved for later use, just as easily and just as effectively as can "real" experiences. Truth be told, they're just as real!

It's a pretty wonderful ability, isn't it? I assure you that you really don't want to lose it! How can it possibly be explained, neurologically? It turns out that it is—as we've pointed out earlier—nothing special, for that wonderfully plastic brain of yours. Remember that you have actually massively constructed and heavily exercised this capacity to control the flow of mental information in memory association, from one thing to another to another, and you use such control to initiate any movement or response. You have already built many hundreds of thousands or millions of threads that can move those memories forward in thought, or carry them forward to action.

You may not have noticed that when you wrote your name with a pencil in cursive in your mind that you took roughly the same time to achieve it as you would have if you had actually written it. In the same way, psychologists have shown that if you imagine a particular map, then imagine moving from one place on the map to another place, the time it takes you—like the musical conductor rehearsing the symphony in her mind—is almost identical to the time it would take in "real" life with the map in front of you. The reason is simple: the machinery that controls this flow in time in off-line thought is the very same machinery that controls flow rates in your on-line processing of "what's happening."

All of your off-line mental operations are crucially dependent, of course, on the quality of your brain plasticity-acquired memory stores. If you don't keep your memory garden weeded, and if you don't add the right fertilizer every so often, all of the operations in thought and voluntary action control that stem from your memory resources will unavoidably follow them to ruin. Pretty soon, like those typical 80-year-olds that we mentioned earlier, your encyclopedia can be recorded in just a few volumes and will use only the most simplistic explanations expressed in the most common language.

Just as our thoughts can move, so, too, do our "old thoughts"—our serially re-constructed episodic, long-term memories. Interestingly, they are also hung on the same reliable framework of place and time that defines our immediate, real-world perceptions and actions. We now know that a deep brain structure called the hippocampus and the brain areas that feed it play an important role in recording and replaying our long-term memories, including what can be remarkably complicated serial episodes drawn out of our pasts. The hippocampus receives information about "what's happening" from most of the rest of our forebrain machinery. It converts that information from the worlds of vision, sound, feeling, smell, and movement to generate multidimensional episodic memories that play out like a movie on a framework of "real" place, and "real" time. We lose this crucial ability if we injure this highest-level memory machinery, or as it degenerates on a path toward Alzheimer's.

The good news: most people have an intact hippocampus. The bad news: it slowly shrinks from our peak-performance epoch, when it is the size of your thumb, heading down to the size of the distal two segments of your little finger, if you're a typical person and if you live long enough. This is just another of a long list of instances in which smaller is not better!

For further explanations and extensive references and citations
related to the information in this chapter, please visit
www.soft-wired.com/ref/ch16

17

A TYPICAL DAY IN THE LIFE OF A BRAIN

Perceiving, Recognizing, Recording, Pondering, Understanding, Responding

Even after reading this far, you probably still do not have a clear understanding of how the rules of plasticity are actually played out in your two- to three-pound brain. That's hardly surprising, because modern brain science is incredibly complicated. Several hundred thousand scientists like me have spent their entire professional lives studying this subject. They've written millions of scientific reports about it. Frankly, I'm often confused myself! Instead of going into any gory details, it might be useful to provide a brief and admittedly sketchy report on how a brain like yours operates. We're going to do that by looking in on my friend George's brain, as it operates across a typical day. There are many trillions of electrical transactions occurring in his and your brain every day that you're alive, and I don't think that you'll want to hear about all of them! I'll necessarily paint a picture of George's brain story with an embarrassingly superficial brush that will only bounce along to color just a few highlights.

Before I wake George up and start his day, while his brain is still idling quietly in slumber, let me tell you just a little bit more about his neurological machinery. Complex electrical information arriving from and delivered back to his body is being shipped on several million ultra-thin "wires" from his sensory, motor, and "autonomic" nerves and lower brain centers up to that very big bulb, the forebrain, which occupies most of his skull. That great hub is under continuous bombardment by millions of sensory "inputs," even while George is sleeping (which he is now doing). In proportion to body size, no mammal has a larger forebrain than we humans. The forebrain is comprised of a great assembly of special subcortical brain regions that feed and support the

95

analysis and action functions of about a hundred distinctive functional zones in the highly folded cerebral cortex that covers its convoluted bulbar surfaces.

That cerebral cortex is a layered structure only about an eighth of an inch in thickness, about two and a half square feet in area. Imagine a thin dinner plate a little less than two feet in diameter and you have the right idea. Within this thin sheet, there are about the same number of nerve cells as there are stars in the Milky Way—about 20 billion of the brain's total of about 80 billion. Those electrically active nerve cells (neurons) are backed up by an intimately associated support staff of glial cells; there are just as many of these housekeeping and otherwise-assistive elements as there are neurons in the cerebral cortex. We'll focus on this highest-level cortical machinery as we tell your brain's tale because it accounts for a large part of your learning, memory, thought-control, and action-control abilities.

The cerebral cortex can be thought of as a vast computer of a special type—of course very different, as we have earlier discussed, from the computer on your desk at home. The cortex is quite uniform in its appearance and organization, and is born without its specific programs (its specific detailed architecture) in place. At the same time its general operating layout and processes have been inherited, and its more than 100 different functional cortical zones (called "cortical fields" or "cortical areas" by 19th century anatomists), each accounting for different aspects of perception or memory or thinking or emotion or action control, have somewhat predictable roles in the brain. Again, it is the specific details of the wiring within and between each of these distinctive functional cortical zones that are substantially "up for grabs" as we plastically specialize their operations through experience and learning.

Unlike the computer on your desk, the cortex operates massively as what computer geeks call a "parallel processor." With this parallel processing you can "walk and chew gum at the same time"—that is, do many things at once. If the cerebral cortex and its supporting structures is the Master Computer, those 100 cortical zones can be thought of as 100 Very Big Computers (VBCs) assigned to different brain sub-tasks, all operating simultaneously ("in parallel"). Those 100 VBCs are, in turn, comprised of about 350 million (or so) "microcomputers." These "microcomputers" are the cooperating nerve cell "teams" that change their team memberships and the extents to which they work in unison in learning, as I have earlier described. Each team extracts and applies their own little bits of information to contribute to the operations of their Very Big Computer (that one-of-100 VBCs to which they belong) and the Master Computer's (the brain's, your) perceptions, thoughts, moods, or actions.

Neurons within each one of these several hundred million little "microcomputers" are complexly interconnected by hundreds of yards of fine, interconnecting processes of nerve cells that you can think of as fine electrical "wires." The fine processes from the sending ("transmitting") side of neurons are called "axons." When I've talked earlier about the staged increase in the insulation on the brain's wires, I've been talking about the insulation coatings on these axonal transmission lines. Axons ship information long and short distances across the brain. Dense bundles of these insulated wires comprise the brain's "white matter." (The cerebral cortex and large, deeper clusters of nerve cells are the brain's "gray matter.") These wires carry information from every brain cell and every brain location—most often to many other brain cells at the same and very often to other (even far-distant) locations. Very elaborately branched tree-like processes from the "receiving" side of nerve cells are called "dendrites."

There is around a third of a mile of fine filamentous processes in each cooperating nerve cell team—which adds up to a lot of complex electrical wiring, especially when you realize that each team, on the average, is a barely visible piece of human flesh that is a fraction of the size of a grain of rice. If we add up the lengths of all of those little "wires" interconnecting all of the neurons in just the forebrain in a young adult, they could easily extend to the moon and back!

How does this complex machinery learn about—and reliably and enduringly encode and record—"representations" of the things of the world? How does it control the actions of the brain? The connections between nerve cells in these vast, incredibly densely packed "cerebral cortical networks" are achieved through small terminal structures called "synapses." There are roughly 1,000 to 10,000 synapses terminating on each cortical nerve cell (neuron). Estimates of the numbers of these synapses in the mature human brain range between about 60 trillion to about 240 trillion. These are very big numbers. Counting to a hundred trillion starting right this second would take you more than 30 million years. If you insist on sleeping, make that 45 million years.

While most of your neurons were born in fetal or early infant life and last for a lifetime, these synaptic connections between neurons are temporary and their coupling strengths are highly modifiable (aka plastic). No synapse can live forever; they all eventually wear out and are subject to a slow, ongoing biological replacement. More importantly, at least most of them are actively "plastic": they can be easily driven by experience or learning to strengthen and to multiply, or to weaken, and to die off. Scientists have documented these plastic changes in many ways and places. A former research fellow of mine now directing a large research center at the University of Kansas Medical Center, Randolph Nudo, taught an adult animal to retrieve small food pellets in a way that required difficult, dexterous finger control. His team estimated that four of every ten interconnections in the small hand-control cortical zone of these animals were altered physically in ways that signified millions of local changes in synaptic strengths paralleling the animal acquiring this very simple skill. Thus, in dramatic contradistinction with the "hard-wired" computer on your desk, the detailed wiring of your own very special personal computer (PC) is very significantly <u>physically</u> different today from yesterday (unless you've spent the intervening 24 hours in a coma.) If you're making any progress in life and learning—which you are, by attempting to read and understand this chapter—your PC will be very significantly different again tomorrow!

I'm sorry that I have to abruptly interrupt this rather fascinating description of the marvels of your brain, because I can see that George is becoming restless, and I just saw his eyes open, then close again as he rolled over on his mattress and pillow. And yes, I can see that he is now gradually waking up. In his brain, the slower, rhythmic beating of his brain responses are showing more higher-speed agitation. His physical movements and eye openings are slowly increasing the levels of a chemical that is flowing out of his brain's dimmer switch. It has begun to flood his brain with that alerting chemical noradrenaline, and as his brain lights go on and begin to brighten, his brain begins to sense the discomfort in his leg and back in its body sense VBCs, the musty smells in its olfactory VBCs, and that nasty, fusty flavor in its taste VBCs, and the ringing alarm clock at his bedside and his dog scratching himself noisily on the carpet next to his bed in his hearing VBCs. It does all that by opening the gates at the brain level just below his cerebral cortex to allow the information from the senses to break through his slumber to generate sharp, locally coordinated activities in the waking parts of his cerebral cortex. It takes the right kind of local teamwork in the brain to elevate his sensations above all of the background chatter in that sleepy brain.

All of these early first-perceived events are predicted by the brain because they repeatedly occur in this place at this time in George's mornings. None of them are being recorded in a strong way on his permanent memory record because they are still not very securely generating coordinated responses, he is only half-attending to them, and his recording switches are only turned up to the "barely ON" position.

As George sits up on the side of the bed, he swoons a little. He didn't use to do that, but now his balance organ and the feelings and the silent information coming back from his body just do not as strongly or as rapidly inform his vascular system that he's in need of those quick, major adjustments in blood flow that are necessary for pumping enough arterial blood all of the way uphill to his now-elevated head. "Remind me," his brain tells itself "to sit up more slowly tomorrow morning." George's brain probably did not record this message to itself very strongly, and it probably won't remember it tomorrow.

George's movements to the bathroom, his first stop every morning, are slow. They didn't used to be so poky, and he didn't used to be so stiff on his first morning stroll. And he didn't used to have to begin every morning by going almost immediately to the bathroom. His brain has gotten into the habit of demanding that he undertake this task without delay. The pressure from his bladder and abdomen is signaling to his brain that the situation is mildly urgent, and it just won't wait very long. The small pleasure and the relief that comes from the act of urination justifies George's pit stop—now, at an older age, as a progressively more-exaggerated habit, burned in by repetition through the processes of brain plasticity that are not always operating in his favor. George has made a habit of blaming all of aches and pains and urgencies and clumsiness on his body. His brain knows better.

It is not until George grabs his toothbrush and discovers that the toothpaste is missing that something more significant happens in his brain. The surprise of its being missing results in a complex pulse of the alerting stimulating chemical noradrenaline. Clarion alarms are telling his brain that "something surprising has happened. We'd better figure out if it's important, or not." A second neuromodulatory area has contributed acetylcholine to that pulse. By its release, it's telling almost the entire forebrain that "something important may happen soon. Be open for brain change."

The pulse of noradrenaline differentially brightens the thinking-and-action control levels of George's brain in the areas just behind his forehead. Those areas, larger in humans than in almost all other mammals, haven't been doing too much up to this point in his morning. As they awaken, they engage the machinery that predicts, from memory, where that toothpaste is likely to be, and they control the head, eyes, body, and hand in the search, moving from the more likely to the less likely places, making quite a few still-less-likely "short stops" because they're easy to check out along the way. In cooperation with a deeper structure in the brain, the hippocampus, and with the engagement of other VBCs in the cerebral cortex, they support a clear image of that toothpaste tube in working memory. George (his brain) knows what to look for. His search-strategy control machinery has long ago mastered this skill by past searching for lost toothbrushes, keys, cars, beetles, children, dogs, scissors, baseballs, books, hammers, pencils, and thousands of other things. As the brain progresses in its search, George's thoughts are moving across those "maps" of toothpaste locations (one of thousands of stored frames of reference that are almost always available) in his brain. Again, George's hippocampus is buzzing brightly as he moves in time and space in his mind. As George searches, his brain moves his eyes from place to place to place, predicting what should be here and there and there and not here and not there and not over there, looking for the anomaly that could be his toothpaste. His brain has spent a lot of time painfully constructing those mapped relationships and those predictable things in the bathroom (and in many other similar) landscape(s) in the earlier history of his brain. There are quite a number of times almost every day that he puts these systematically mapped environmental location reconstructions to good use!

Because George is failing in his toothpaste hunt, his brain guides his stream of consciousness to move across other associative paths for following alternative paths in thought or action control. Some of them are pretty crazy. "Brush my teeth with soap. What am I telling me?! Me, myself, and I must be joking!" George leads his self (actor-referenced action-control networks) to cut off the search, and brush his teeth without toothpaste. "Not entirely satisfactory," the emotional side of his brain tells his brain, with the dark emotional coloring that comes from failure or disappointment. As he brings the brush to his teeth to engage those elegant tooth-by-tooth and tongue representations that take up such a large territory in the tactile sense machinery in his brain (so elegant that he, or you, can detect even the smallest of surprises in your mouths), he misses the usual, sharply remembered sweet taste, and the lubricating feel of the creamy toothpaste. Again, the unusual taste and feel is a surprise to him, even though if he had thought about it, it should not have been. That surprise has occurred because his brain knows that brushing his teeth has been sweet (and he likes sweet) and it knows how that toothpaste has lubricated the brushing on those many preceding mornings. The brain just can't help itself from predicting that sweet lubrication. Once again, the unfilled prediction turns on the alerting chemical machinery, which gives George a moment of special attention—as always applies for the unexpected. Perhaps you have not been aware that you and George have been constructed to be surprise junkies. TV, videogame, and movie producers are very well aware of that fact!

Because the conditions were right (George was paying attention; he had some degree of focus on the task; surprising things happened) George's brain will be able to remember his search for his toothpaste, and will be able to reconstruct the serial details of that search for most of the hours of the day ahead of him. By tomorrow, more uncertainty about those details will begin to creep in to what will soon be misty memories. That does not mean that these experiences won't be used by George's brain for any one of a million possible reasons a month or a year or ten years from now. He and you have recorded many millions of little seemingly trivial events that your brains can use. It's just that only one in a thousand of them will ever come up in your mind in a sharp-replay form. Many, many more are in there, but they're lying low! Nor will his or your brain ever ask you if it can put this background information to use. In fact, it can't tell itself exactly when that is going to happen.

As George begins the very automatic daily routine of shaving—a task that he can achieve almost without thinking about it at all, his stream of consciousness moves him to imagine that Marian has probably done something with the toothpaste. George's brain is generating a somewhat ghost-like vision of her moving the toothpaste, but that little movie flickering in his brain doesn't get so far as to show him exactly where she might have put it. His associative machinery is having a wonderful time guiding the flow across the Never Never Land of his thoughts. George's emotional VBCs inform him that he's a little angry about this possible crime that he has decided his mate just might have committed. "It's silly to get mad about that," his brain tells his brain. "Probably just forgot." The idea flashes through George's mind that he has been becoming a little more forgetful himself. "Is it possible that I did something with the toothpaste? Could I have carried it out of the bathroom absentmindedly?" His brain harrumphs itself, at the fleeting thought. A pang of guilt rises in his thoughts as he remembers that Marian is helping her sister and isn't here to defend herself or tell him where he can find the damn toothpaste. He (his brain) relies on her so much for so many little things like this. It is as if Marian and her memory were an extension of him. And so she is.

As George completes dressing, his movements are again automatic, requiring almost no attention or thought. His brain plasticity switches for all that postural change and touching and visual-motor control are again "barely ON." Those automatic pilot-controlled activities don't interest the dressing parts of George's brain enough to allow much behaviorally driven change. It wasn't always that way, of course—but it has been a long time since George had to pay any close attention to the details of putting on his shirt or tying his shoes.

While George's brain was on automatic pilot for dressing, it was very much turned on by other emergent thoughts. His episodic memories carried him to a reconstruction of this day's appointments and obligations, each one predicting the next in serial order, each one leading him to associations about how he will handle each task. George organizes those higher executive-self levels of his anterior brain operations in this way multiple times every day, and if he or you added up all your hours spent practicing your forward-planning mental skills over the course of a year they would number in the thousands. That's good news, because the loss of these abilities marks the advent of very big problems at an older age. George's brain is still pretty good at iteratively refining, then hardening, today's work plan into an action plan so that he's pretty sure (and a dopamine-evoked smile affirms) that his day will be well spent.

While George was doing all of that mental planning and memory embedding, he casually turned on his radio and listened to a touching story on National Public Radio about the unusually complicated struggles of a young injured soldier trying so hard to overcome his PTSD and re-establish a happy civilian life but whose utter failure ended with a tragic suicide. Even as George was mostly thinking about his schedule, his emotional VBCs were strongly engaged and he could feel his eyes becoming a little watery. "Gee," his brain tells itself. "I didn't used to be so weepy." Earlier in his life there was more neurological control over his moments of sadness, but as he ages, the sophistication and power of this control is slowly weakening.

With the radio on, George's listening and language brain have come alive; as the story drew him in, he lost track of schedule-planning for the moment, and his more acute listening (he closely followed every word) arose from that amplification of listening that comes from those moments when what George or you are hearing is really important to you (your brain). George's recording machinery was saying "SAVE THIS ONE." His tear-producing and emotional machinery had obviously also come online. Because he was actively listening, and because this story was important for him, he'll remember it throughout the day—and for many days or even years to come. Each time he tells this tale, it will be a little different. Those differences will be incorporated into a slowly evolving memory of "what happened." Still, each retelling will again likely bring a tear to his eye because the sadness is an integral part of the remembering and its emotional reinforcement will embed these slowly evolving memories all the more deeply in his brain.

The text messages start coming in about this time every morning, because George's family, friends, and his team at work know that they can rely on him being up and about by this early hour. His attention comes to be more and more strongly divided between his morning routine and background listening, because he always responds to the texts. The familiar chime leads him to grab the phone that is literally attached to him, burned into his personhood in his brain as a direct, almost-always-present extension of the person that he is. It responds for George at high speed, as his fingers command, because his brain knows that it is now a part of him. If George or you thought about it, you (your brains) might be just a little bit ashamed that the two of you have given things like your smart phone that much power, so close to your neurological core. But on the other hand, maybe the two of you (your brains) need all the help they can get!

By the time that George walks down the stairs toward the kitchen, shoes in hand, his muscles and joints are loosening up, and he feels livelier in his step. He has to pay a little more attention on the steps because he slipped and nearly fell on the stairs three weeks ago and knows it could happen again. It was surprising that when he began to fall, he didn't seem to be able to catch himself to prevent it. Because it was important and frightened him just a little, his brain remembered to remember. This thought flew through his mind in the middle of the lower flight of stairs, just before he heard then sensed his dog Doug hopping down the stairs next to him. As Doug brushed George's leg, his brain sent out the orders to stiffen that leg, just in time to prevent a fall. "Gee," George (his brain) said to himself, "it's a good thing I was thinking about being careful. If I had been just a little bit slower, or not thinking about the possibility of falling, I'd have fallen again." His brain congratulated itself for both reminding itself about that earlier fall and predicting in the nick of time that Doug might bump his leg. An associated worry coupling his emotional VBCs to his action VBCs to his memory VBCs flashed through his mind (brain). "Next year, I may be slower."

Because it sensed another small movement, George's brain was alerted to tell his head and eyes to move to the site of the disturbance, where George saw his happy dog against the predicted backdrop of this tiny little part of George's World waiting for him at the bottom of the stairs, tail wagging and eager for his expected pat and expected breakfast.

George's brain predicting the danger of Doug moving past him on the stair was remarkable because that exact situation had never occurred before in his experience. Fortunately, as in many such instances, his brain had enough other related experiences that led to falling or almost falling or to just maintaining his balance to enable an immediate "guess" (estimation, prediction) about how his brain had to control the actions in his body to prevent a tumble.

George used to walk Doug every morning, but George's anxious brain tells him that he just doesn't have time for it. Doug can spend his usual half hour in the backyard, and George will let him back into the house before he leaves for work. He's grown very fond of Doug—almost as if Doug is a part of George. It is as if when George sees him, he feels Doug is him. His brain agrees. So does Doug's.

George's preparation and eating of breakfast could almost be done in his sleep. He knows exactly where everything is in his kitchen. The bread, butter, peanut butter, muesli, milk, coffee, grinder, water. He's pretty sure that he could successfully make and eat breakfast in total darkness! The map of his kitchen in his brain is pretty much cast in concrete. George's brain would be strongly alerted if there were anything in his breakfast life out of place. While he didn't used to be so regimented, it sure makes life easier.

As George sits down with his toast and muesli, he picks up the controller and hits the red button to turn on his television. There's almost nothing on in the early morning that really interests him, but it's on, anyway. He just likes the sound and is attracted to the voices. Once in awhile he hears something interesting. Once in awhile there's some important news about what's happening in town, or in the world. He quickly searches all of those older familiar TV addresses, seeking stimulation. Once again, like you, George is constructed to be just a little bit of a thrill seeker.

As George walks out of the back of the kitchen to his garage, he feels even livelier in his step. When he backs his car out onto the driveway and into the bright sunshine, it is a pleasant surprise to see that it is so nice in the city this morning. His dopamine cells are achieving their warming effects, but of course they've done that in the same circumstances a thousand times before, which is one reason why they're so good at cheering him up.

I can carry this story forward through George's daily commute, workday, dinner and dancing or movie, or I can carry him away into postprandial passion, have him go the opera or to the bathroom two or three more times, have George watch an hour or two or five of TV, play FarmVille and buy a pig or two, play a fast game of Tiddlywinks or maybe Boggle, spend 20 minutes on his treadmill moving it up to 7.5 degrees, have him brush his teeth again and maybe this time lose his toothbrush, undress himself, then fall into fitful sleep—but enough is enough. I suspect that you're getting bored by my story, and just don't want to hear any more about George. Perhaps your brain is bored by my story as well. I also suspect that I have done a truly terrible job of describing any real day that your brain has. Your (brain's) days are, I'm sure, more interesting, more exciting, more important, more alive.

Then, on the other hand, even while that is perhaps all true, perhaps your brain (you) is (are) also just a little bit (and maybe very) bored by its (your) typical-day brain story—which is one reason why I led you through this rather agonizing tale to make you think about what your brain may have to be putting up with, on a regular basis. Does my little narrative have any analogies or any resemblance to your daily story?! It's time for you and me to discuss how you are actually using your brain, and what may be good, and not good about it!

I also had a second more important reason. Did you happen to notice that, through the course of every moment in the morning in George's (and your) day—and in fact, throughout each and every entire day, for every single second without a single break, George's (your) brain was (is) HIM (you). Lots of people have the notion that they are something quite different from their brain. They're not.

For further explanations and extensive references and citations related to the information in this chapter, please visit www.soft-wired.com/ref/ch17

18

TRANSFORMATION, PART DEUX

How Improving the "Little Things" Can Change the Course of a Life

Kenneth worked as an electrician from the time that he arrived as an immigrant from his native Belize more than 30 years ago. He enjoyed the variety of everyday challenges that he faced on his job. His chosen line of work had provided a good living for his family, and he was proud of his professional abilities. When the State of California passed a new law that required every electrician to pass an exam for a license, it came as a humiliating shock to Kenneth that he did not pass. After his first failure, with his pride and his family's welfare on the line, Kenneth threw himself into a serious period of study. He found that it was difficult to focus and to absorb all of the new information required for licensure now that he was in his early 60s. Still, he was surprised when he failed the exam a second time—and now very worried, when another round of study culminated in a third failure. "Three strikes and you're out," seemed to be Kenneth's fate.

Kenneth heard about neuroplasticity-based training programs that were being offered for free in a wonderful adult school program run by the Los Angeles Unified School District. He had achieved almost everything important in his life through a dedicated program of self-improvement. It made sense to him that a restoration of order in his world might require neurological self-improvement.

Kenneth did restore order. Through intensive training using one of the brain training programs that we created, he experienced large-scale personal improvements in his "brightness," his memory, and his ability to absorb his study materials. In Kenneth's words, "I could remember a whole lot of things I couldn't remember before. It made me more focused and my mind didn't stray anymore." Kenneth passed the relicensing exam, on his fourth try, with flying colors.

Catherine was married to a domineering husband for nearly 50 years before his passing left her—so everyone thought—somewhat adrift. Like many lasses in Ireland in those days, her father had taken her out of school as a young girl because school cost a little, and "because it wasn't going to come to much use anyway." Catherine was being raised to be a dedicated mother and subservient helpmate who would be expected to leave all decisions to the future ruler of her household. With her husband gone these many years later, her children and friends were pretty certain that she would struggle with the management of her own affairs. She could hardly read—indeed, had never read a book in her life. She had never progressed in school past simple arithmetic and had never had any say in family finances—and could not be imagined to be able to take care of them. Because she'd been homebound, Catherine also did not seem likely to be able to find the help that she needed to overcome these deficiencies—which she had endured, after all, since childhood.

When Catherine's son Art showed up unexpectedly to surprise his mother with a visit, he noticed that his mother had a phonics reading training program booklet sitting on her worktable. His mom, now over 70, was teaching herself to read! Art later discovered that the women at the Senior Center in her community had all been talking enthusiastically about a book that they had all shared that Catherine had actually managed to read all by herself. His mom thought that since her friends seemed to be so interested in this novel, now that she was making a little progress in her reading, she should see if she could read it for herself. Once she had the book in hand, the librarian at the Center told her several times that she should return it—but because reading was slow for her and she had to look up quite a few words, she invented reasons why it would be just a little bit longer before she could manage to bring it back to the Center.

In time, Catherine was very proud of finally completing that first novel—then, in rapid succession, its two sequels. I laughed a little when Art told me that his wonderful mother, who had never read a book in her life rather accidentally began a life of reading by struggling through the sex- and passion-filled Fifty Shades of Grey—then, at an accelerating pace, those two follow-on volumes filled with lots more shady gray prose! For Catherine, what a wonder it all was, to finally have the world of literature open to her—and perhaps all the more fun, for a woman steeped in Irish traditionalism, to begin her literary adventures with a work of such astounding and eye-popping fiction!

Now that Catherine has emerged out into the open air, all on her own, it turns out that she can take care of herself. She can grow and flourish without much help needed, thank you. Because Catherine's brain, like yours, is plastic.

For Maggie, the early signs of HIV-associated cognitive impairment were subtle. "I was having trouble with word retrieval. First it was people I didn't know," Maggie explains as she describes how she forgot the names of famous politicians and actors. "As time went on, it began to be people I knew, like colleagues," and eventually close friends—all this while still in her 40s, not an age at which one normally expects to experience this level of cognitive decline.

Maggie is like many millions of other individuals who carry additional neurological burdens stemming from other medical problems that she has had to endure in her own special passage through life. She had been able to function at a high level while working as an HIV-positive attorney for years, but over nearly two passing decades, the low-grade brain infection that is a part of the HIV syndrome produced increasingly more significant cognitive deficits.

Maggie found herself omitting letters when writing simple words—which is an important problem for a hard-working practicing attorney. "Between my brain and my hands things would just kind of go missing." HIV was particularly affecting the quality of her handwriting and the accuracy of her hearing and remembering. "I was having more problems hearing in noisy environments like restaurants, and was having to ask people to repeat themselves." These cognitive challenges slowly eroded her confidence and pushed her to withdraw from social and public speaking activities related to her job and her AIDS activism.

To make a difficult situation worse, Maggie's father suffered a stroke, affecting his memory. At this time, Maggie heard about the neuroplasticity-based training that we had developed and validated, and thought that it might be a source of help for both her and her dad. She was right.

"My word recall improved dramatically. I feel that I can speak fluently again. I can do public speaking and I have been. I can speak with friends and family without embarrassment." Maggie saw her handwriting improve significantly, exactly eleven days after starting the program. Her father—who had seen his own penmanship suffer after the stroke—independently noticed his own improvements on that same 11th day of training.

Father and daughter trained together, motivating each other to stick with the program. "He would come home exhausted and still insist on going upstairs to spend his hour working at the Brain Fitness Program (now a part of BrainHQ), Maggie fondly recalls of her father's dedication.[5] His hard work paid off. By the end of their training, her father's memory and other important cognitive abilities had very significantly improved, along with other practical everyday skills that contributed to his sustained independence.

Of course this "brain correction" training also helped Maggie with her memory, word recall, and handwriting—but perhaps the greatest benefit was something more abstract. "Just in general I feel more confident. I feel that once more I'm just more involved in life."

There are millions of Kenneths and Catherines and Maggies and dads out there in the wider world who have begun to understand that they are not doomed to their "fate," that their brain is plastic, subject to correction and re-strengthening, far out into the future. It's not just about holding on, as a survivor and witness of your progressive decline, or your loss of fortune. It's about growing again, in all of the many small ways that make all the difference, for sustaining—or, if necessary, for re-claiming a better life.

For further explanations and extensive references and citations related to the information in this chapter, please visit www.soft-wired.com/ref/ch18

---

[5] See www.brainhq.com.

PART FOUR:
THE BRAIN IN RETREAT

19

LOSING GROUND, JUST BY HAVING A BIRTHDAY

Our Abilities Change as We Age

Near the end of his second term, President Reagan told one of the White House physicians that old joke that we've all heard: "Doctor, there are three problems that I have that I simply must tell you about. The first is that I'm having trouble with my memory. The second...uh...uh...uh...." Four years later, his doctors informed him that he was "mildly cognitively impaired," a milestone on the path to Alzheimer's.

While his doctors never thought that President Reagan had the formal symptoms that would qualify him for receiving the "dementia" label while he was in the White House, there is little doubt that this exceptional man was slowing down mentally through this period of his life. That's hardly surprising, for a man who was sworn into office early in his eighth decade. If you're a little bit on the older side of life's Great Divide, you may have some inkling about what I'm talking about—because there is a very high likelihood that you have also begun to slow down. If you're a typical person between the ages of 40 and 100, you're probably slower than you used to be in your movements, mental actions, and reactions, even if you practice a lot—and even more so, if you don't. That's why golf has a "Senior Tour." It's why Satchel Paige was the only player in baseball history to still have a job on a professional team beyond the age of 50. It's why people in their 70s just can't click that button fast enough to win much money on "Jeopardy."

When we think of slowing down, we often think about it in terms of movement control, but it also applies to our abilities to perceive, remember, think, and reason. In listening, we become less accurate at following a conversation, but our errors are especially notable when we listen to someone talking rapidly or when we have to come up with a quick riposte in conversation. Have you noticed that you have more trouble understanding a fast-talking child? That occurs because in most older individuals the brain's accurate responding to rapidly successive sounds is slowly deteriorating. Beyond about age 60, the average person begins to make errors in what they hear—even when the speech of others in a conversation is delivered at a normal discourse rate. As the years pass, everyone seems to be either mumbling or talking too fast.

If we study this in a research lab, we find that if speech is artificially sped up just a little, most people over 60 begin to have many errors in understanding. If words that are twice as fast as normal are presented in random order, most 70- to 90-year-olds understand almost none of them. If we present the same kinds of fast speech to people between age 20 and 40, under exactly the same conditions, they understand almost everything.

As the years pass by, those sound details are represented within your more sluggish brain in a progressively cruder way. Your brain therefore struggles to record and remember speech because it's relying on neurological patterns that aren't as clear. Needless to say, it's a lot easier for you to continue to operate with aplomb in your social world if you understand and remember conversations.

Older people also move their eyes less frequently, spending more time staring at a particular point in space between relatively infrequent eye movements. As we age, we require stronger or more intriguing events to occur in front of us to drive our eyes to move to look at them. As with older listeners, the information that the older observer acquires to create a complete brain picture of a visual object or scene is sparser than for a younger person. After all, that older brain is often recording a scene from information gathered from a single staring glance, while in the same time period a bright-eyed young individual is bouncing their eyes around the same scene to take five or more "snapshots" of it. Who do you think is going to do a better job of scene analysis and recognition? Who is going to do a better job of recording—again, remembering—all of things that are happening in that scene?

An older individual is also slower in using information that they receive—or have previously received—in their mental operations. That is particularly crucial because information in the present moment gives us such an important advantage for accurately judging what is happening in the next moment. This slowing in ongoing predictive biasing is easily demonstrated by simply timing how long it takes you to report on what it is that you have just seen or heard. The brain has to "know" what it has just seen or heard or felt if it is going to be able to control the biasing for the next thing (or things) it expects to occur. The older you are, the longer that "knowing" takes. I could set my stop-watch to record the time it takes you to solve a new puzzle or answer a new problem presented to you. On average, the older you are, the longer you'll take. Indeed, if you set the timer on almost any mental or motor task, older means slower.

Scientists have also related brain speed to measures of human "intelligence." If you challenge me with a problem or a puzzle, how many possible solutions can I evaluate, mentally, in any given span of time? 60-year-old me, on the average, can only get partway down the list of possible solutions that can be considered by 30-year-old me. By age 70, the average person will mentally evaluate fewer than half the options that the average 30-year-old will come up with.

In parallel with your slowing down, you lose accuracy, precision, and fluency in your behavioral operations. When these losses apply for your sensory-guided movements, ("I used to be able to thread a needle more quickly!" "Why do I keep dropping things?!" "Why did I trip over that?" "Why did I pronounce that word so strangely?" "Why did I almost fall, just getting up out of that chair?") it is natural to blame your blundering and tottering on your physical body. Your body is, indeed, partly to blame. But your body has a sneaky partner residing within your spine and skull who is also not working up to its old standards. Moreover, your loss of agility applies to your thinking just as it applies to your movements, and the losses in mental agility can't be blamed on a creaky body.

As we've pointed out, the loss of both speed and accuracy contributes importantly to that poorer ability to remember. Remember names. Remember poems or jokes or gossip. Remember just how you are supposed to tie that necktie in a four-in-hand bow. Remember what you do after you add the eggs. Remember where you left your keys. Remember the name of your son-in-law's sister. Remember how to spell "amnesia."

Your problems with memory actually cover a broad range of important everyday skills. Remembering that (formerly automatic) set of things that you need to do to perform some task that you used to be able to do unthinkingly, literally with your eyes closed. Remembering something that you just did, or just heard, or just saw. Remembering who did what. Remembering something that was said a few minutes ago, or that you are sure you saw yesterday. Remembering the details of that plot twist in that novel. Remembering something that happened a year ago, or 50 years ago. Remembering what came fourth, after you completed the third instruction. Remembering all of the things that you used to remember belonged together, like the members of that school committee, or the names of those nice people down at the garage, or the aisle where you can find the vinegar, or the depth, spacing, and fertilizer that were recommended for planting those dahlia bulbs.

If you want to conduct an experiment on yourself to illustrate how good your memory might be in another 20 years, pick up your portable radio and walk around all day with ear buds or a headset listening to a position on the radio dial where all you hear is continuous hissing noises, then see how well you can remember the things that you've heard on that day. That's pretty much what your brain will be doing! Adding that noise in the background as you are listening—or degrading the clarity of your vision by simply turning the lights down low—results in an immediate decline in your ability to remember what you've just heard or seen. In an older brain, that "noise" is growing from the inside, since the way the brain encodes what you hear becomes less precise— fuzzier—over time. You don't actually hear this growing, internal chatter; it lives in the machinery itself. Still, that growing internal "noise" has exactly the same effect for degrading your ability to record sights and sounds as would turning on that hissing radio or dimming down the lights. As this internal infernal hiss grows, for your memory operations it's a classic case of "garbage in, garbage out."

It's important to understand that this kind of fault cannot be completely overcome by simply practicing how to remember. An individual with progressive memory loss hasn't forgotten how to remember. Their problem is that their brain is struggling to permanently record things because they are now encoded in a sparse, imprecise, error-ridden manner. You can practice endlessly, trying to record such neurological trash as a memory, or to retrieve it from the trash heap. From the perspective of really improving your memory, doing all of those crossword or Sudoku puzzles, or working at all of those "memory games" on the Internet, is an exercise in enthusiastically kicking a dead horse!

When you're older, distracting events interfere more strongly with what you're trying to hear, see, or do. Accurate listening requires your selective attention. The older brain is much less effective at suppressing the millions of interferences that come from modern, busy, cacophonous, or visually chaotic environments. This problem arises in part because the "signal"—the brain's activities that represent the thing you're trying to remember or do—is growing weaker. Unfortunately, to compound the problem, the ability of the brain to quiet down the activities generated by all of those unimportant distractors occurring in your listening or visual or body sense environment (and from intrusive meanderings and thoughts springing from the brain itself) is also far weaker than in the younger brain.

A famous McGill University neuroscientist, Herbert Jaspers, who was still actively working as a scientist into his 10th decade of life, once explained this noisy-brain problem to me. "Mike," he said with his wonderful raspy voice topped by a wry smile, "when you get on a bus to go to some particular destination, it's not just a matter of getting on the right bus. You also have to be careful not to get on all of those wrong buses." What Dr. Jaspers meant, of course, is that your brain has the difficult task of continuously suppressing all of the irrelevant or incorrect alternatives that could result in error, or could disrupt your neurological flow. Older brains have increasing trouble quieting down those numerous sources of interference and interruption that contaminate our modern social (and internal thought) environments.

An engineer would say that an older brain's "signal-to-noise ratio" is poorer than normal. A poor signal-to-noise ratio on your TV results in a fuzzier picture and a background buzz on the audio. Interestingly, scientists have recently shown that an individual's growing problems in keeping those disruptions under control are a main source of your failures in "real world" remembering. Unfortunately, in all likelihood, your modern "real world"—both within and outside your skull—is almost certainly very noisy.

When you're older, you simply see and hear and feel less. You are probably unaware of it, but as the years pass by, you actually take in a progressively narrower slice of the world in front of you in your active vision. By the time you're 80 years old, the portion of the horizon in front of you that you reliably see and report on is only about half the view of the world that you were taking in at age 20. A typical 80-year-old is walking through the world with blinders on! Not surprisingly, many things that happen just outside of that field of view that could be important to you—like another car moving toward you at the intersection that is on a trajectory to hit your car—no longer attracts your eyes to them. This limitation in what you can see as you grow older is a major reason why many people over about 60 just cannot drive as safely. This is especially true for driving at high speeds, because even if an older driver sees an impending hazard, their visually guided decisions and actions can be so much slower. It's also true of driving in the dark, because older people have additional problems with glare due to age-related changes in the eyeballs.

Most people from middle age onward (and millions of youngsters as well) also suffer hearing losses that are exacerbated, as in vision, by compensatory changes that occur in their plastic brains. Hearing losses in aging are substantially a product of our modern cultural environments. Scientists who have studied people that spend quiet, drug-free lives in remote natural environments are more likely to have more-intact hearing designed to last as long as the rest of their physical bodies. By contrast, our ears are under continuous acoustic and pharmacological assault, and almost all of us suffer partly self-inflicted hearing losses in middle and older age.

Our inner ears also house a crucial set of sensory organs that control our balance. This very fragile sensory apparatus is also almost certain to deteriorate with the passing years. As we'll talk about later, the brain resources supporting your posture and balance are pathetically under-exercised in we modern humans. As this system slowly degrades, the brain may find itself more dependent on the slower information from the visual system and from the body senses that tell you that you're unstable, beginning to stumble, or about to fall down.

Given a hearing loss, we can amplify our bad hearing with a hearing aid. We can compensate for a hardening and rounding lens in our eyeballs by wearing corrective glasses for close-up vision. We can even ask our ophthalmologist to replace a fissured lens or a misshapen or cloudy cornea. Unfortunately, we have no simple repair strategy for improving the information that our older, deteriorating balance organ might be delivering to our brain. And unfortunately, we have no equivalently simple solution for improving the quality of the poorer information delivered to our brain from our aging skin, muscles, joints, or innards.

It might be just a little bit discouraging for you to read that you are slowly losing your senses. But remember that you can take control of your brain. A key to your adult life is to use your brain to get the most out of the still-powerful information that it is still continually receiving from your several worlds. We'll consider how you can achieve that later in this narrative.

As brain processes slow down, as precision declines and memory and thinking slowly degrade, the brain's range of functional operations slowly simplifies. That occurs partly because some things were wonderfully and richly mastered (or powerfully repugnant or distressing) in earlier life, and are therefore more easily sustained in your brain than the millions of things that your brain was involved in that were not so important or so frequent. Some momentous stories will never be forgotten. Some skills, performed a million times, are burned into your brain, and available to you for the long haul. They still work for you, and as a consequence, you can use them in your reduced life (with a danger that you can use them to the extent that they can bore the devil out of your family and friends). It is partly because your brain has a long history of relatively confident or mildly insecure or mildly fearful or mildly irritable or mildly acquisitive or mildly lots of other things that can be exaggerated in a brain that is telling itself to simplify, simplify, simplify. I'll tell you stories about people that have gone down this path of progressive simplification in chapter 21. These strong forces that contribute to these changes in an older life are called "negative learning." In chapter 24, I'll also describe a number of other ways in which negative learning exacerbates functional decline, so you can better understand what might be done about it.

What are some specific things that can occur as a consequence of negative learning?

1. Life can become less joyful. On the average, older people do not feel as alert, bright, or vital when they get up each morning. It just isn't as much fun to travel, go to the symphony, or walk the dog as it used to be. Attention often flags. There can be lots of days when you just don't feel like doing anything.

2. Older individuals learn more slowly. They can still learn even very complicated things, but it can be harder to get started, and more difficult to sustain the effort. Progress is usually a little (and sometimes a lot) slower.

3. Your brain just won't let you go to sleep ... and to make matters worse, it also does a poor job of waking you up. That is because the machinery that controls your learning and remembering also controls sleep initiation and waking—and because it is being neglected and under-utilized, it is slowly dying. Growing insomnia and slower awakenings of brain and body are indicators that your brain is losing its way as the healthy learning machine that it was designed to be.

4. Many older people turn inward, toward a more egocentric life, and this can have strong negative learning consequences on your life and your brain. Like other remembered associations, an older person may find the forces of attachment to other individuals weakening. That more self-centered older person can slowly withdraw from the ongoing challenges of operating in the social and physical world, limiting themselves to the safer havens of the deeply embedded and the extremely familiar. Nothing in your brain is as frequently referenced or as familiar as yourself. If you feel yourself sliding down this egocentric slope just a little, it might be time for you to think about what you might do to throw out that anchor!

As you read this tale of functional decline, and later, in chapters 22 through 26, about the neurological mayhem that causes it, I want you to remember that there are bright sides of getting older. Older people have a proportionally more extensive history. They have a greater measure of understanding that comes from longer, richer experiences. They are better integrators, in large part because they have a lot more knowledge and experience to integrate. They really are, each one of them (us), a living documentarian of their (our) fast-changing slice of cultural history. It is the rare 20-year-old who has any particularly interesting stories to tell.

But most of all, we must never forget that given brain plasticity, mature individuals have the capacity to rejuvenate their brains. With rejuvenation, we all have the potential to more richly exploit those extensively recorded assets that can only come with age. What if older people could recover the speed and efficiency of a younger brain, while maintaining full access to those rich stores of knowledge and experience? That would allow your brain, if or when you are older, to operate with special powers.

For further explanations and extensive references and citations
related to the information in this chapter, please visit
www.soft-wired.com/ref/ch19

20

IT'S NOT JUST ABOUT GROWING OLDER

Many Other Things Can Contribute to Neurological Struggles

"I'm not old, not impaired, and I'm not losing it yet. Actually, I'm operating at a pretty high level. You're making me read all about brains getting older and feeble. This has nothing to do with me!"

My reply would be that this has to do with everyone, no matter their age. Your brain is plastic. You—anyone, everyone—can be stronger, more capable, more competent, and more intelligent. Plasticity is all about getting the most out of each brain, for the potential betterment of each human life, at any time throughout that life. Just as importantly, while you may well be at the top of your game today, in possession of a brain that needs little assistance or work, you won't be able to sustain your lofty position as you age unless you take this science to heart. To have any real hope of making the most out of your life throughout its course, it is important to understand how things can be expected to change as you age, and how many unexpected things have the potential to set you back.

"I've been reading this book because I have a problem that I want some clear advice about dealing with, and all YOU want to write about is what goes wrong when you grow older. That's not MY problem. I (or someone I care about) have (has) a history of _____ [Fill in the blank]. When are you going to talk about that?"

Almost every day, I receive emails like this from people all over the world. They detail their own neurological or psychiatric problems, or describe the problem of a parent, child, mate, cousin, or patient about whom they care deeply. The majority of people on this planet either have or will have a brain problem that has or will require medical treatment. Almost all of us love someone who now falls within this class.

Almost all of you who are near-peak brain performers today can look forward to a more-troubled neurological history, or will have to confront neurological problems when they plague someone that you deeply care about.

While I have focused on normally aging brains, it is important for you to understand that the sources of neurological difficulties and the principles of corrective brain plasticity apply just as much for an individual with a brain that is wounded, traumatized, developmentally impaired, environmentally impoverished or twisted, poisoned, infected, addicted, depressed, obsessed, phobic, anxious, attentionally disordered, oxygen-starved, psychotic—or any one of a thousand other brain-based maladies—as it does for aging or, for that matter, for growing the potential of even a currently very high-performing brain.

In every case of chronic impairment, illness, or disease, plastic changes in the brain contribute strongly to the expressions of the illness, as the brain struggles to sustain control under difficult conditions.

In every case, brain systems are disabled or distorted in ways that manifest condition-specific faults that distinguish one specific brain disorder from any other.

In every case, no simple chemical or physical manipulation can be expected to cure the malfunctioning brain all on its own, if "cure" means the full restoration of normal brain function.

In every case, plasticity-driven brain remodeling is a necessary component for any real and complete correction to occur, to the extent for which correction is possible.

To understand how plastic changes in a brain contribute to the expressions of neurological and psychiatric dysfunction and disease, it is useful to describe several real examples, because brain injuries come in a myriad of different real forms.

Let's begin with JD, a young Texan who was born with inherited weaknesses that ultimately led his brain down an alternative developmental path, which was expressed as a moderately severe form of autism.[6] In Texas, if it's at all possible, a child like JD is mainstreamed in a regular classroom. No teacher in JD's elementary school was happy to hear that he had been assigned to her class. JD was hell on wheels, not just because he couldn't sit still and was a continuous source

---

[6] You should understand that the neurological changes and behavioral expressions in these examples only very superficially describe the true picture that applies for these conditions. For some addition explanation and references, see www.soft-wired.com/ref/ch20.

of distracting noises and movement. JD was also physically strong, which made him difficult to manage during his relatively frequent classroom and playground tantrums. JD often spun off into highly distracting periods of "stimming" marked by elaborate and repetitive hand and body movements and low but audible moaning sounds. How could his classmates stay on task with JD anywhere in the room, obsessively stimming away? How could they be expected to suppress their grins and teasing when JD answered so many questions by just repeating the question? Perhaps most frustrating of all, while JD was reasonably verbal and for significant periods was under control, he made little real progress in his studies, seeming to struggle to understand and to read, unable to keep on task long enough to do very much meaningful work in class, only rarely completing his homework assignments.

It wasn't JD's fault. For a child with a learning impairment, it's never their fault. JD was born with a brain with multiple genetic faults that ganged up to frustrate the brain's advance in infancy and childhood in ways that would normally provide it with a sharp resolution of the details of what he heard or saw or felt. His brain did not develop the normal powers of oral speech reception, language usage and language-based cognition. It simply did not evolve the complex machinery that provides the basis of social understanding, and JD's social control remained in a near-infantile state. His brain had never developed a normal level of selective attention control. Without attention focus, the brain can't create a stable, predictable, reliable world model, because it has to build a world not from "what goes with what" but from "what goes with just about everything." JD's stream of consciousness ran forward as a broad, shallow stream that covered almost the entire landscape of predictive possibility as it flowed haltingly forward in thought, and in the control of his actions.

At the age of eight, JD's future prospects could hardly be described as promising. In the usual course of the life of a kid with his level of autism, it is likely that he will be bounced out of public school not too far into his future, especially when he advances into those socially complex and sexually charged teenage years. It would be lucky if JD's abilities could grow to the point where he could be expected to have a happy independent life. His mother and father are already deeply worried that they will have to provide for him to the end of their days, and despite their limited resources, will have to put some plan in place to help him when they no longer can. There are several million JDs out there living among us, largely invisible.

Randy's childhood was very different from JD's. Randy had always been a high achiever in school, in his profession, and in life. In his 50s, one particularly strong passion of Randy's was bicycle racing. He had realized that he had special capacities of endurance, and exploited that gift by competing in 24-hour cycling events in which distance traveled determined the victor. He also particularly enjoyed competing in long distance events that had to be completed over a large part of a day (or two) without a break. Randy was one of the best in the world in his age bracket in these demanding competitions. Before long, no one was surprised when he outperformed individuals of all ages in the events that he chose to compete in.

Of course maintaining your abilities at a high level in any sport requires a serious schedule of conditioning. On his bike, on one of his daily workout stints, Randy's competitive career—and in many respects Randy's happy life—literally came to a crashing halt. When he was struck by a passing vehicle, a crushing blow to Randy's brain jeopardized almost everything that had been important to him in life, put everything at risk over the course of a few seconds.

A mild-to-moderate diffuse traumatic brain injury (TBI) like Randy's occurs almost two million times a year in the United States. A large proportion of individuals have had at least one TBI in their lives, many as participants in contact sports like football, hockey, or soccer. More than a third of the approximately two million soldiers and marines who have served in Iraq and Afghanistan have come back home with a TBI. TBIs are often devastatingly life-altering.

Your brain's wiring system is made up of millions of axons. Each axon contains stiff microtubules—infinitesimally small tubes—that deliver nutrients and chemicals to synapses at their terminal ends. Those nutrients and chemicals are necessary for sustaining those trillions of points of connection in the brain. A sharp blow to the head can generate broadly distributed damage to these microtubules, resulting in the loss of many millions of brain connections over the subsequent hours. To appreciate the scale of damage we're talking about, these tubules are only about 1/3000th as thick as a human hair, and a traumatic brain injury can induce millions of these minuscule points of damage. Think of your brain, in a flash, as being shot full of millions of tiny little holes and you begin to get the picture. When the brain is exposed to a blast injury, as has occurred several million times for military and civilian populations during the wars in Iraq and

Afghanistan, brain cells look like they have gone through a hurricane. As the pressure wave sweeps across the brain, cells are distorted and jumbled, with the largest neurons especially affected because they present such large, elaborate surfaces to the pressure wave, like the largest and leafiest trees shuddering and bending and twisting under the full force of a gale. Physical breaks and tears damage large numbers of these neurons, and strip off some of their most important synaptic connection points.

In this kind of injury, where there may be no direct penetrating wounding or no large zone that is completely destroyed as, for example, after a stroke, aneurysm, brain-penetrating wound, or brain surgery, losses usually affect almost all brain functions. While no ability is completely lost, just about everything the brain-injured person does is degraded at least little—or, as in Randy's case, a lot.

Randy's neurologist told him that life was going to be very different for him. He would have to live with his impairments for the rest of his days. Cycling was just one of a long list of things that he'd have to forget about, for the duration.

The decline in Greg's abilities occurred far more slowly. His life's work and all of the things that he loved to do in his personal and social life were dependent on a sharp mind and on his keen memory and wit. It was obvious that they were now failing him. Greg knew the cause. He'd carried the burden of his HIV infection with him for more than 20 years. The drug treatments that have saved so many lives had controlled the infection in his body, but unfortunately, those miracle drugs don't entirely block the progression of the viral infection in the brain. Slowly, year by year, nerve cells are impaired, simplify, and die in the brains of HIV-AIDS patients like Greg. The connections in some brain areas are reduced, which strongly impacts language, memory, reasoning, and fine motor control. Behavioral and mental losses directly parallel the deterioration of these damaged brain areas.

"I had short-term memory problems so that I couldn't remember lists of things beyond two or three," he recalled. "I had difficulty remembering names and getting a hold of common words." His ability to work with numbers was diminished to the point that it became impossible for him to continue in his medical research career. Coming up with the names for the simplest of everyday objects became a challenge for Greg. "I would say, "you know the thing you put bread in and it makes it brown"?" meaning, a toaster. For a person well-versed in neuroscience like Greg, this class of problem was expected, given the intense bloom of the virus seen in brain scans of the superior temporal lobe and frontal cortical areas that contribute so importantly to language function and memory.

Forgetting the word for "toaster" was only a minor part of Greg's problems. He was also particularly frustrated by a sharp decline in his manual dexterity. We know that HIV infections are particularly virulent in brain areas that contribute to the control of hand movements and other fine movements, so this expression of loss was, again, hardly surprising. With shaking and clumsy hands, Greg struggled to unlock the front door of his house, or to insert his key in his car's ignition. Greg felt that his useful life was slipping away from him. Uncorrected, it probably was.

You might note that Greg's brain, like Randy's, had suffered from diffuse damage. When the brain is infected from any cause—as it can be after West Nile virus, Lyme disease, meningitis, or any number of brain-invading infections—damage is widespread. Restoration requires that the affected person work in a rich variety of ways to achieve a recovery that's anything close to complete.

Similarly diffuse injury occurs after the brain is poisoned. Lead, cadmium, arsenic, PCBs (polychlorinated biphenyls), PBDEs (their polybrominated cousins often used as flame retardants), and many hundreds of other brain-damaging chemicals out there in our physical environments poison the brains of millions of individuals exposed to them at brain-damaging levels. We also voluntarily ingest damaging chemicals when we submit to (usually absolutely necessary) regimes of chemotherapeutic drugs to kill our cancers, or to psychoactive drugs to bring our pathologically unstable brains back under control. Such involuntary and voluntary brain poisoning can simultaneously save— while in more subtle ways degrade—the lives of millions.

Brain damage is also diffuse after oxygen deprivation that can occur because of arteriosclerosis, in an accident because of strangulation, near-drowning, or a collapsed lung, in cardiac or vascular surgery, and in many other scenarios.

No matter the cause, diffuse brain damage leaves the brain in a position very much like that "diffusely degraded" brain of the aging patient. Almost every quality that I attribute to the aging brain also applies for brain change in the face of a neurological problem, and for recovery, because that can only occur through brain plasticity that re-strengthens and functionally restores degraded brain processes.

Allie's problems also grew slowly within her. She knew they were coming. She knew that they would get worse. She expected to lose control. She anticipated her insanity.

Allie had begun to recognize the seriousness of the problem during her freshman year in high school, when she knew that the thoughts in her brain could not possibly be real, even though her brain told her that they were. She knew that she bore the stigmata when her friends just stared at her shaking their heads, or when she overhead the "Allspice" nickname that they used.

Still, Allie did well in school, and was especially strong in math. She had succeeded in college on a high level, completing her degree in mathematics and accepting a fellowship at the University of California to pursue her Ph.D. She couldn't tell you now just when she slipped into madness. She knew that she was receiving information directly from a special source that provided powerful new insights into how to correctly translate information that was hidden in the words that other people actually said or wrote. People didn't realize that this hidden information was there, and they didn't realize who was making them say or write it. It was a secret that Allie was eager to tell others about, because she understood the code and could interpret the hidden messages. It was enormously frustrating for her that no one seemed to take her seriously or understand what she was talking about.

Allie was sent to the doctor repeatedly, and each time acquiesced and accepted whatever pills they prescribed, but she certainly didn't believe that she needed them, and she was certainly not going to take them. She wished she could make the doctors understand that they were the ones who needed medicine; they were in danger, not her.

At age 25, Allie's life had run off the rails. Her professional opportunities were in shambles. She was socially isolated, friendless, and she believed she was estranged from her family despite the fact that they were desperately trying to help her. At age 26, a period of hospitalization finally convinced her to take her prescribed medications—and when she did, her madness was controlled to the extent that she could live a more peaceful life. But while the demons were quieter, they were hardly gone, and she was still seriously cognitively and socially impaired. Successful employment, independence, and personal happiness were remote and unlikely. Allie's brain, after all, had been undergoing progressive changes for many years. Elementary brain processes controlling alertness and learning were now greatly distorted by her disease. Her brain's associative powers were now so weakened that they could not keep track of who did what in her world. "Did I do that?" "Did you do that?" "Did God do that?" she might ask herself. A breakdown in associative memory ultimately breaks down a person's very being; in the words of William Butler Yeats, "Things fall apart. The center cannot hold."[7]

When a person's ability to predict what consequences follows from what act, the social behavioral machinery in the brain is slowly—and ultimately profoundly—damaged. Even the simplest of abilities, like remembering which voice or face goes with which name, or distinguishing a look of amusement from a look of amazement or disgust, can be lost. Brain processes that control complex problem solving are frustrated by a stream of consciousness that moves sporadically and erratically across a broken landscape. Only one of Allie's brainpowers was superhuman: the distortions in her brain machinery powerfully informed her that her own brain was the only truth teller. No one is more certain that their notions are true and profound than the schizophrenic individual whose thoughts are often in utter disarray, and so obviously and utterly untrue to everyone around them.

Almost all neurological conditions that might lead you to the door of a psychologist or psychiatrist arise in a brain in which there is a distortion in the alerting, learning, memory, or fear-avoidance systems of the brain. Every addiction, every obsession, every phobia, every psychosis, every anxiety syndrome, and every depression is marked by such a disturbance. Alas, in each of these classes of illness, there are always plasticity-generated consequences that largely account for complex illness-specific distortions in brain function. While there is a large degree of overlap in these distortions, there is also a panoply of individual changes that account for the behavioral differences between these different patient cohorts. These issues affect many: over 100 million people in the U.S. alone are now being treated, or have been treated, for some disorder in this broad neuropsychiatric class.

Bruce had no problem with problem solving. His problem was, in some respects, almost comical when compared with Allie's. Bruce

---

[7] … the self-descriptive title of a wonderful and chilling first person account of schizophrenia by Elyn R. Saks.

simply couldn't see most of the left side of the world out in front of him—and he didn't know that he couldn't see it. When Bruce had his stroke, his ability to understand speech and the movement of his right arm and hand were affected, and he was worried that he would not be able to operate effectively with those impairments. Fortunately, he was able to overcome those problems, and in his mind, there was nothing wrong with him anymore. However, for everyone else in Bruce's life—including his neurologist and his wife Mariana—the true picture was not so rosy.

Like Bruce, the majority of individuals who have a stroke, aneurysm, or penetrating wound affecting the left side of the brain experience the inability to process and perceive stimuli on the right side of the body or environment, a condition known as hemi-spatial neglect. You hear a lot about stroke patients who can't talk, move, or think, but not so much about an equally large population who just can't see nearly half of the world in front of them. While this deficit may not seem so important to you, if it remains uncorrected it will strongly impact Bruce's life. He cannot safely drive. He will literally stumble his way through life. He'll look at the full scene, centering his vision in the appropriate way for his former visual abilities, but now, the left half of every book, every TV image, every object, every person—including Bruce himself when he looks down at himself or looks at himself in the mirror—will go missing.

Focal brain damage destroys brain tissue; no brain plasticity can replace Bruce's missing flesh. But it can replace function. In every other condition that we have described, the brain is still physically intact, ready to be engaged to self-correct, through appropriate intensive retraining. After a brain lesion like Bruce's, recovery still requires plastic remodeling, but in a case like his, the goal must be to recover function using neurological machinery and strategies that are obviously not (cannot be) critically reliant on the now-missing brain tissue.

My colleagues and I have studied recovery in animal models, for example, by documenting the machinery that the brain was using to master a specific complex task, then completely destroying a source of crucial information necessary to achieve that task mastery. Following its stroke, such an animal now utterly failed at the task. We then trained the animal intensely, and in the end, saw that it could perform the task just as well as before the stroke. In our model, we saw that the animal actually expanded and elaborated an alternative, undamaged, initially modest source of information as a consequence of re-training. This newly elaborated source of sensory feedback now provided the crucial information for the machinery of the brain that had earlier been provided by the now-destroyed brain areas. If there is any alternative way to master lost behaviors after focal brain damage—as is frequently the case—recovery can potentially be achieved by remodeling the brain, via intensive training, in ways that exploit both its other sources of crucial information, and its inherent plasticity.

When a patient has a brain wound, their brain naturally struggles to perform its formerly important behaviors, even while its performance is distorted. If I've lost control of my hand and arm, I will instinctively try to reposition my hand through any still-intact abilities. If I can still control my shoulder movements, I'll naturally try to use those movements to control the positioning of my hand. Working hard at this every day, I might ultimately become very adept at that particular ability. Looking into my brain, I would see a very well-connected and active brain area now dedicated to controlling fine shoulder position. Brain plasticity in this respect can be a kind of trap for the brain-wounded individual. Because plasticity processes are competitive, I may have unknowingly established unequal competition between an intact shoulder-controlling brain and a grossly weakened arm- and hand-controlling brain. If I really want to recover more normal function, I have to give the weakened competitors—in this example, my arm and hand—every advantage. Insights like this one, coming from brain plasticity science, have had a significant impact on rehabilitation success in the clinic. The processes underlying brain plasticity are richly informative about which specific re-training strategies are going to provide the strongest and most complete recoveries.

It should also be pointed out that when the brain is wounded, as it has been in millions of our fellow citizens, changes occur within the brain far beyond the extent of the physical lesion. The brain can actually change the balance of its wiring pretty dramatically when key sources of influence are lost. Many areas and functions remote to the wound may be weakened, disconnected, or disturbed by the loss of the contributions from the "dead zone." A stroke patient may not fully appreciate this broader weakening, as their brain struggles to do its best with the new burdens that it now has. If the individual with brain damage wants to be whole again, these more general problems should also be addressed, along with the more obvious deficits that would tie them, without recovery, to their wheelchair or armchair.

While these descriptions are almost inexcusably superficial and only relate to a small fraction of the problems that can actually challenge a brain, I hope that they help you understand that brain plasticity science relates very directly to understanding the expressions of—and the behavioral consequences that arise from—<u>any</u> neurological dysfunction, damage, or disease. It is also a core science for understanding how the problems that arise in dysfunctional, damaged, or diseased brains might be most effectively addressed on the path to the most complete recovery. When a brain is poisoned, diffusely damaged, oxygen-starved, or genetically weakened, it's challenged in ways that are akin to the changes induced in natural aging. Distracting internal chatter grows and clouds normal brain processes. The brain slows down in all of its operations. It becomes less accurate and organized, struggles to record information with normal accuracy, and struggles to maintain sophisticated control of its actions. Almost every aspect of brain function is negatively affected.

When the brain machinery that controls alertness, learning-induced change, or fear-avoidance is distorted, it biases the changes to more seriously impact social, emotional, and decision-making processes in the brain. Because these distortions influence virtually all of the great processing systems of the brain, they ultimately impact almost every aspect of brain function. As in the case of diffuse brain injury, the impact can be brain-wide. There is a lot to fix if any true cure is to be achieved.

When the brain incurs damage that results in the death or removal of brain tissue, a person commonly experiences severe problems that are more limited in scope. A person may not be able to speak, understand speech, move their limbs, control impulses, see a complete world, or stop their body from moving. These problems must be corrected with an understanding that correction requires that the brain be trained to change itself so that, insofar as possible, new ways are found to perform old tricks. At the same time, no large lesion in a brain limits its effects to any single or simple class of problems. Many brain processes, including many that are relatively remote to the wound, can be affected. A true cure has to acknowledge these extended secondary losses.

I ask you to reflect, now, on any one of these conditions and ask yourself a simple question: "If I have such a brain (or if I love someone whose life is frustrated by such a problem), how can I (they) ever expect to be well again?" My answer: "Through a lot of hard work, using your (their) brain plasticity to your (their) very best advantage, and driving faculty after faculty back as far as possible in an improving, stronger, corrective direction." There is no question that you can be stronger, more capable, and possibly recover and even grow past your old limits.

While the specific focus of how you spend your time in retraining your brain will be different in each of these conditions, the overall principles are the same. For those of you reading this book with this kind of history (or for those of you who love someone who has it), all of the information and advice that I offer for normally aging individuals applies especially strongly for you.

In case you are wondering, JD, Randy, Greg, Allie, and Bruce are all real people who came to our research and development laboratories with these problems. All have done their best to engage their plastic brains to drive them to a stronger place, in a recovering direction. JD was trained over a period of months, and has continued training to sustain positive impacts over his development. He ultimately overcame the most severe problems that seemed so strongly to limit his future prospects. When training brought his autism under control, he exhibited high intelligence, and was identified as one of the better students in his grammar school class. All that stimming, all those tantrums, and all of that hurtful teasing are now largely a part of fading older memories. While his improvements were greater than in most autistic children with his level of impairment, they were well documented by extensive behavioral testing. I have no doubt that JD now has a bright future.

With Randy's determined spirit, he has completely recovered his life by training his injured brain. He has used our brain training programs extensively for nearly three years, and in his words, "I am now virtually 100% recovered from my injury. From all accounts, I have beaten the odds. I have been back at work for over two years now. I am back to competitive cycling. I have resumed a productive life. I am now doing many of the things that my neurologist said I would most likely have to give up .... I volunteer time to help support the survivors of TBIs and their caregivers and regularly recommend the programs to those in the midst of recovery and seeking activities that will help."

Greg spent several years trying to find a strategy that could help him slow down the HIV-induced changes that threatened his future. After a series of failed efforts to help himself get stronger, he heard about our training programs in what proved for him to be a life-changing moment. Within a few weeks of focused brain training, in Greg's words, "Everything was going up, up, up!" Greg was surprised by the "tremendous positive benefits" of this more-targeted training. His memory returned. His dexterity recovered. His faculty with numbers was regained. His overall cognitive abilities improved. Clinical assessments verified that his broad-scale recovery was for real, a finding that has been extended in a controlled study with other HIV-AIDS patients. All of the cognitive and physical impairments that he had previously experienced had either evaporated, or had been significantly reduced in magnitude. Needless to say, these gains have had a big impact on Greg's daily life. In his words: "I'm back."

Allie also participated in one of our brain-training studies, and the cognitive gains combined with the positive effects of her medicine have allowed her to acquire and maintain a job and live independently. While her recovery is far from complete, she can live a fuller life again, back in her community. She has made several friends and is closer once again with her mother and father. They are no longer living in dread that she will end up on the street, where it can be so difficult if not impossible to help her.

Bruce can see it all again. As a subject in an experimental trial that has been conducted by our research group, he recovered ear-to-ear vision after only a few hours of computer-based training. He's really happy that he's gotten his driver's license back. By all clinical evidence, he also has his life back. Bruce only grudgingly appreciates this: remember, he was never able to admit to himself that there was anything wrong in the first place! But his wife, family, and doctor are exceedingly happy with the results.

When the brain is injured, when something is gnawing at it, or when a behavioral distortion is growing within it, it changes plastically to accommodate these losses and distortions. It does not just sit still. All through that progression of negative changes, that plasticity could have been redirected to drive the brain in an improving, corrective, strengthening, and empowering direction.

The sooner you get those healing changes in motion, the better. If you're a little late in getting your own brain moving up rather than down, remember that wherever the progression of your problem has left you, even after many years of potential negative change, your brain is still plastic. You can always be better.

For further explanations and extensive references and citations
related to the information in this chapter, please visit
www.soft-wired.com/ref/ch20

21

WHEN YOUR BRAIN CHANGES, YOU CHANGE

Older Should Be Wiser, Better, and More Balanced, but Alas...

My paternal grandfather William Theodore Merzenich immigrated to America from Germany at the age of nine. He had almost no further formal schooling in Fairfield, Minnesota, where his family first settled. It might be surprising for you to hear, then, that my Grandpa Bill was one of the best-informed and most intelligent men that I have ever known. For most of his life, he worked as a building contractor and architect. During World War II he worked as a U.S. government building inspector, then as an "Engineer of the Ways" at a shipyard. Grandpa was proud of the several patents he was awarded, and of all of the churches, school buildings, and civic buildings that he contributed to throughout the northwestern United States. He had mastered the basic principles of practical mathematics, so he didn't need a structural engineer to tell him how to carry loads and construct buildings that he believed would last forever, and he could talk at length about the laws and principles of physical science. He had broad knowledge about literature, culture, philosophy, and the world in general. Everything he knew and every ability he had mastered had been acquired through determined self-education, or learned on the job. He seemed to work almost every day of his life on an organized program of intellectual self-improvement.

My grandfather had very high personal standards in his work, and his children, grandchildren, friends, and professional colleagues were expected to live up to them. In his view, there weren't nearly enough other members of the human race that really measured up, a view that became stronger with every passing year. As he got older, Grandpa Bill became more and more certain that he knew the answer to almost everything that was worth knowing. By his 70th birthday, there were a lot of folks in Grandpa's world who he felt had sloppy thinking and loose ethics that were leading them all down the path to ruin. Grandpa was dead certain that these negative human qualities contaminated those he saw as these weaker members of the human race. Lawyers, politicians, other automobile drivers, irresponsible and under-qualified practitioners in the building trades, and citizens who could not control their urges were at the top of his list.

But my Grandpa was not just your ordinary curmudgeon. He invited discussion and arguments from his grandchildren, family, and friends, then laced his side of the discussion with lots of wonderful parables and memorable stories on his way to dialectical victory. He was actually fun to lose an argument to, even when you had heard one of those stories ten times before. Of course he enjoyed arguing in part because making his point and declaring victory were richly rewarding for him. One can understand why people like Grandpa Bill might grow to look forward to such interactions, when their brains tell them that they're never wrong and reward them with a win every time they go into combat!

Meanwhile, Bill's wife, my grandmother Leila, enjoyed gardening, and as she approached her 70th birthday, she was serving as the President of our small town's Garden Club. The wife of the Governor of Oregon decided to do a little politicking by visiting garden clubs across our State, and this rather elegant lady invited herself to come visit our town. The Club decided that my grandmother's garden would be the perfect place for an afternoon tea. For my grandmother, this was a social distinction and honor, and being a kind, hospitable, and agreeable woman (an absolute necessity for putting up with Bill) she quickly consented to this plan. Unfortunately, the Governor of Oregon was, in my grandfather's view "... a G—D— charlatan and hypocrite," and to put an even finer point on it, made sure we all knew that "that dumb s— of a b—'s wife is never setting foot on my property!"

How many times have you seen a forceful, compelling individual exaggerate those characteristics as they grow older, to become a caricature of the more elaborate and more balanced younger person that they once were? At a younger age, Grandpa had a lot more "give" in him. He would never have been this self-centered and insensitive. But by the age of 70, whenever he made a forceful pronouncement like this, he was trapped. His brain would not allow him to back down.

My father and his brothers solved this problem by colluding with their mother. Keeping it all a secret from Bill, they kidnapped him for a day, inventing reasons why he had to leave early in the morning and stay safely away from home all day and into the evening. That gave Grandma Leila time to set up her party and have a nice chat with the Governor's wife about roses and hydrangeas and camellias while everyone enjoyed their tea and crumpets. When Bill came home and discovered this betrayal, he fumed and fussed and stormed about, mad at everyone and everything. I remember that we all avoided him for several days because we just couldn't stand all the hollering.

Almost everyone who knew my grandfather respected him. When he died, friends came from all over to express their admiration and affection. At his wake, I remember a round table in his home piled high with cards dedicating Catholic masses for him in remembrance of his passing. While the older Bill Merzenich had undergone a slow transformation to a more simplified and more extreme form of the younger Bill, as he aged, he richly compensated for being so almighty sure of himself along his journey by bringing humor, laughter, caring, and other forms of generosity into his interactions with the people that he admired. If he liked you, if he thought you were a person of honor, and if he thought that you were trying your hardest or doing your best, you could count on his loyal friendship and support.

Strength of conviction in our beliefs is surprisingly weakly correlated with empirical truths. It grows in our brains through reinforcement, but those reinforcing messages from outside sources or from our own brains do not have to be factual or sensible. The assignment of certainty is a kind of neurological trick your brain plays on you throughout your life. You have probably encountered many people who have extremely strong convictions about a particular belief, despite being totally ignorant of the particulars of that belief. All of us are wrong about some of our most firmly held beliefs.

For example, my Grandpa believed that he was one of the world's best drivers. In reality, he was a terrible driver. He'd lost an eye in an accident on a job site in his 60s and had no depth perception. He never knew exactly where his car was in relation to the cars around him. As he harangued all of those other drivers who he thought were out of control, the fault was almost always his. This is not too surprising, because a recent poll of drivers showed that nine of ten people believe they are "above average" at driving—when obviously, it's mathematically impossible for that to be true. Then again, another poll found that 87% of people considered themselves "above average" in math skills.

It also bears mentioning that Governor Hatfield was actually pretty well-liked, and considering him "a G—D— charlatan" and "a dumb s— of a b—" was definitely a minority view.

The story of Grandpa Bill is meant to illustrate that it would be better if you don't evolve into a caricature of your younger self. As a person ages, it would be better if you could sustain your full social adulthood, and not slip back in the direction of more-childish ego-centrism. I'll talk about strategies to at least slow down those processes of regression in later chapters.

My wife Diane's mother, Marge, made near-perfect pies every time. She had mastered this and many other kitchen arts, and for many years her grateful son-in-law could always look forward to an exquisite, warm piece of pie with a wonderfully flaky, buttery crust and a sugary fruit or custard or nut filling. Thousands of wonderful pies later, how could this dear lady come to believe that her pie-making skills—along with other domestic skills and abilities—were abandoning her? Why did she come to believe that pie making was just too difficult or too chancy? Why didn't those tens of thousands of historic compliments override her slowly growing insecurities?

Confidence comes from success. When we're younger, we're right more often than wrong. We remember more often than we forget. We meet our standards for the majority of the things that we attempt, because, after all, our capabilities are still expanding and improving as we're still establishing those standards. When we're older, we may find ourselves being wrong more often than not. We're not terribly old before we come to the point at which we may forget more than we remember. But, just a little unfortunately, we still remember those high standards from earlier life, and realize that as the years go by we may not be so reliably achieving them. Marge's growing insecurities were another form of a regression back to a more childlike view of life. As her brain grew in its reliability in that earlier epoch of life, her self-confident actions grew with it. As her neurological strengths waned as the years passed by at an older age, so, too, did her confidence.

If you no longer believe that you can make a really good pie, you make less of them, and like Marge, you may always announce to your guests and to your brain that you can no longer make good ones. Unfortunately, as with Marge, your brain will be listening! Over and over and over again, your brain is reminded that pie-making is, for you, a more and more imperfect, and eventually, a hopeless endeavor.

In my Grandpa's case, we see the brain of a person who, brimming with secure confidence, has transformed into an intellectual bully toward the later years of his life. He had constructed a life out of always being right, and his brain was driven to distortion by all of those rewarding, self-announced victories. In striking contrast, Marge's brain slowly toted up every little error, and utterly lost the confident control of its actions. The result was an increasingly defensive person whose brain had been driven to distortion by all of its (her) little self-proclaimed defeats. Both individuals, in very different ways, grew into exaggerated caricatures of their younger selves.

My older brother Ed became friends with an elderly widow (we'll call her Mrs. Thompson) who lived alone in an old farmstead near his home. He would often stop by to check in on her to make sure that she was okay and to offer her a little companionship. Not infrequently, she had a physical chore for him to do, or an errand to run. As she got older, he helped her with her bookkeeping and taxes, and at some point she asked him, solemnly, if he would act as the executor of her estate after she died. When that untoward day finally came, Ed found that her living relatives were not much interested in the things in her country home, perhaps, in part, because most of her possessions had long ago been buried under mountains of useless junk. It may surprise you to learn that about 20% of older people develop into hoarders like Mrs. Thompson. She had saved everything that had come into her house for many years. Several large truckloads of garbage had to be removed from her home before it could be sold to settle her estate.

I have known several other serious packrats who just can't see their way to putting any piece of mail or newspaper in the trash, or who must always have stocks of food in place just in case all the stores are closed for two or three years, or who can never say "no" to that beautiful jewelry offered on the TV shopping network, or who fill their basement or their entire house with so many things that they "might have use for some day" that it would be absolutely impossible for them to find any one of them, if and when that need should finally arise.

This obsession with "stuff" is a product of a brain that has been rewarded hundreds of thousands of times across a lifetime by the accumulated rewards that come from acquisition. In the end, the brain can come to need those rewards, every bit as much as it needs food itself.

Monica has developed into a rather remarkable "neat freak" in her older years. A minor scratch led her to completely refinish her hardwood floors, because no other remedy would produce the perfect result that she required. No speck of dirt or dust can escape her eagle eye. Her little dog lives a perilous life in her home, because even with her nails neatly trimmed and her hair clipped on a regular basis, any small upset can result in a big upset for Monica. Her fanaticism about cleanliness extends to her spotless porch or garage—where eating on the floor would be a sanitary step up from eating on most people's dining room table—and out to her perfect yard, where every blade of grass is taught to obey. If a garden tool or appliance shows the smallest blemish or hint of rust or wear on it, it has to be replaced by something that is, in her words, "more reliable."

Monica's husband's main assignment in life is to help keep their home pristine. Nothing can be out of place. No dust or defect can be visible. Believe me when I tell you that this is a full-time job for all concerned! Her husband Gerald has complained that while he is well past retirement age, he can't quit working because he'd have to spend all day long with her, responding to her many complaints about the shortcomings of his domestic assistance.

Mrs. Thompson and Monica have both became caricatures of their younger selves, in the older phase of life. One lives with extreme clutter; the second can only live in that parallel universe of extreme anti-clutter. Both transformed, older people are exaggerated extensions of the shopper-acquirer and stylish-shipshape individuals that they could both take some justifiable pride in, in earlier life. Both grew their tendencies, through brain plasticity extending over many years of "practice" into something beyond mere passion. A growing eccentricity or obsession that distorts who you are, and that can slowly redefine you to almost everyone else's considerable annoyance, is a natural, common outcome in our plastic brains—unless, of course, you work hard to avoid it.

I have an old friend who dates back to my young college days (we'll call him "Ned") who has led a grand life, especially in his own mind. Ned has told me many fascinating stories about his life. He's at the center of almost every story, or rides out into the sunset to save the day, if that's what's required for matters in his tales to come to a good conclusion. Most of these stories, I am quite convinced, have at least some basis in truth. Most also have been improved substantially over Ned's lifetime.

Ned truly believes that these revised, enhanced versions of his life accurately describe it. I'm a living witness who can affirm that they don't.

Several years ago, I wrote an autobiography with the goal of writing a history of my family for the benefit of my children, siblings, nieces and nephews, and cousins. I told many stories of my childhood just as I remembered them. When I had completed the book, I sent it to my five siblings, asking each of them for their comments. Each one of them had memories of different specific family events that were not identical to my own. Some of these discrepancies were impossible to resolve; we simply remembered the same events differently.

Who's surprised? One of the first things the 19th century scientists who initiated the serious study of memory discovered was its deceitful nature! After struggling for several days to make amendments to my family story as my siblings' comments came in, I disposed of the most contentious differences by telling them that if they disagreed with my version, they'd just have to write their own books.

You should not be too surprised to hear that this problem grows as time passes. Scientists have shown that repeated stories slowly change over time, and that with that re-telling, new versions come to be accepted by the re-teller as the truth. Many changes creep into our recollections as we grow older. As our neurology simplifies in the last decades of life, the oeuvre of many older people shrinks so they start telling a small number of stories over and over. Because of the high rate of repetition, those stories can become reduced in form, and can become more stereotyped. Research has shown that older people are actually pretty good at remembering who told them a story, but they are terrible at remembering who they have already told a story to. This works against them on two levels: on the one hand, an older Larry or Sally Ann is in danger of telling the same story over and over again to the same person without realizing it, to their potentially substantial annoyance; on the other hand, if someone tells either one of them the same story over and over, they remember that they've already heard it and are likely to get just a little annoyed themselves.

Ned has always been a storyteller, and as a man with mild, lifelong self-esteem issues, he has always been a larger-than-life action hero in the tales he tells. But as he ages, the older Ned has evolved into an even more exaggerated caricature of his younger self, making him almost unbearably egocentric and narcissistic. What if Ned found some new stories to tell? As he ages, that would be a great blessing for his relatives and friends.

My wife Diane belongs to a book club that has been together for more than 30 years. Her fellow club members are the closest of friends who honor that friendship by elaborately celebrating each member's birthday. That used to be easy to arrange, but as the years have passed, I've noticed from a distance that it has become more challenging. Now, it seems like it takes days just to organize a single birthday event on which everyone can agree. Why do we increase the frequency with which our uncertainties, anxieties, and eccentricities grow in ways that impact our willingness to just say "yes"? Why do we increasingly insist that our own point view is so important, and must be considered? I'm sure you agree that this trait, shared by many older individuals, is actually more childlike on everyone's part.

Now let's imagine that Diane's book club invited my Grandpa Bill, Marge, Mrs. Thompson, Monica, and Ned to join their group. Who knows, maybe they'd be discussing this book! What we can be sure of is that all would have very strong views about what was right or wrong or up or down, over how and where to celebrate that next birthday. All would come to the table with a more egocentric perspective, slowly grown within their plastic brains as a distortion sprung from many, many thousands of self-generated brain rewards or brain punishments. It's a wonder that older individuals can get along at all! In fact, some of them do not. Which leads me to my last story.

My maternal grandmother died at the age of 37, following six of her seven siblings and both of her parents into the grave. In the span of about 20 years, all had succumbed to tuberculosis. Three years after her mother's death, my 16-year-old mom welcomed her new stepmother into her home. Theresa was a vivacious, cheerful woman who met the definition of being "the life of the party" in the most positive way. She was the beloved "Grandma Theresa" for us children. Theresa had been raised in a convent orphanage, and graduated to a life of domestic service before marrying my grandfather. Her marrying a successful widower farmer near her 40th birthday was a way out of a life of service for the wealthy family that she had cooked for, for so many years.

How could this positive, generous woman be consumed, in her later life, by the fear that her loving relatives coveted her home and farm and were actively plotting to take them away from her? How could she come to be so sure that their interests were under-handed, self-serving, and avaricious? How could the old insecurities of a young life arise again, after a period of prosperity and joy, to infect, in exaggerated form, her older life? Her slow transformation manifested a revision back to a reduced and more egocentric person again capable of deep insecurities, jealousy, and bitterness, directed in a childlike way to individuals who loved her deeply.

In the end, like many older individuals, she spent the last years of her life in social isolation, alone, suspicious, continuously anxious, and deeply unhappy. Although she lived past her 90th birthday free of Alzheimer's disease and senility, the distortions in her brain were disastrous. What a great sadness it is to see a brain evolve to that kind of dark place.

All of these stories illustrate the ways in which an aging person may be transformed into a simpler version of their former, younger selves. These changes occur slowly, and as we move backward from the peak performance period of our lives, we continually define ourselves in the present moment. The more socially capable, balanced person that we were 10, 20, or 50 years ago is very difficult for us to reconstruct in our memories. When we remember familiar things—none more familiar than ourselves—we continually update them, day by day, year by year, and decade by decade. Wonderful people like Grandma Theresa or Grandpa Bill or Mrs. Thompson—or even you—can slip almost unawares into a place that it would be much better to avoid.

In chapters 30–37, I will describe some ways that you can help your brain maintain its younger-adult balance and social agility, and help you consider how you can avoid letting it regress back in that more egocentric, childlike direction that it quite naturally wants to slip back into.

For further explanations and extensive references and citations related to the information in this chapter, please visit www.soft-wired.com/ref/ch21

22

MACHINERY IN NEED OF REPAIR

Physical, Chemical, and Functional Changes in Older (or Wounded) Brains

My uncle's brother Dick was, from the time of birth, cognitively impaired. He compensated for this grandly with a wonderful natural friendliness, and we all loved him dearly. Dick was hospitalized after a major automobile accident, and the doctor, noticing that he had head lacerations and was acting pretty oddly in the hospital, did a brain scan. We all received the bad news. "Poor Dick appears to have suffered major brain damage. He will probably never be normal after this accident." My cousin Dan and his wife Greer, who lived nearby, rushed to the hospital to comfort him. Sure enough, it was the same old peculiar Dick that we all loved! It turned out that the scan had revealed a shrunken brain, which was status quo, for our old pal!

Around the age of 60, the brain begins to slowly shrink in volume to be a little more like Dick's brain. Those changes affect the cerebral cortex of the brain, that great mantle of nerve cells that account for much of our "higher" brain function. They reduce the volumes of many important brain structures that support mental operations and movement. And they shrink the brain's white matter, the great network of connections in the brain. Changes are greatest in the areas that affect memory and thought, attention, social behaviors, navigation, and complex actions. These brain areas and the higher-level functions that they account for were the last to fully mature in our juvenile and young-adult lives. As I have noted earlier, these last areas of our brain to come online are the first to go offline as our brain function regresses back in the direction of the chaos that marked its earliest years of life.

Many scientists have focused on age-related changes in an important brain structure known as the hippocampus, as well as the brain areas that feed into it, because this machinery has been shown to play an especially important role in long-term memory, way-finding, and complex mental reconstructions. Indeed, the strength of activation of the hippocampus has been shown to be strongly correlated to an individual's performances on a variety of memory (and other) tasks. As memory declines with age, the hippocampus and related structures undergo progressive shrinkage, and begin to generate correspondingly weaker responses.

Other scientists have focused on age-related changes in the frontal cortex, which is the brain region just behind your forehead. The frontal cortex also contributes importantly to recognition and memory, complex thinking, sophisticated control of attention and actions, and social behaviors. Again, volumes and levels of activation of the relevant areas of the frontal cortex have repeatedly been shown to decline with age. And as with the hippocampus, the differences between older and younger brains are strongly related to the performance differences in memory and other abilities.

Still other scientists have focused on structures that support other important functions like attention, voluntary movement control, language abilities, different dimensions of learning, and so forth. They have been able to document changes in brain activities and size of various related brain structures, and correlate them with differences in expressed ability. You could summarize these many studies by asking a simple question: "Which brain regions controlling which brain functions are negatively affected by age, as indexed by the status of the nerve cells, the complexity of their interconnections, the chemical machinery that supports their actions, and the functions that they support?

The answer? *Almost all of them.*

As we age, in one functional system after another, we lose many nerve cells, and the ones that remain become simplified. Although there is some capacity to replace dying nerve cells, their slow, continuous demise outweighs their possible rate of replacement. Still, with important exceptions noted later in this chapter, the great majority of nerve cells in most brain regions actually survive with surprisingly small net losses in numbers, right up to the time of onset of Alzheimer's disease or to the end of life. However, among those that survive, their interconnections progressively simplify. Imagine the average nerve cell in a young brain as looking something like a majestic, mature oak tree. An older nerve cell looks more like a scraggly, diseased tree that has lost most of its smaller branches.

The formation of connections in a young brain has been described as the construction of a complex transportation system, replete with paths, trails, byways, country roads, streets, and highways. This system sits within, and connects with, an inherited infrastructure of freeways and autobahns. In older age, our road crews are working in reverse. Paths and country roads have been ripped up, and streets and local highways are not being maintained. If we don't do something about it soon, our excursions are going to be limited to going to only the most prominent destinations. To put it another way, that myriad of interconnecting brain wires that once got you to the moon and back may now leave you stranded well short of home, stuck somewhere out in space. Brain shrinkage is largely accounted for not so much by nerve cell loss, but rather, by this progressive simplification of those trillions of interconnections between your nerve cells.

What is happening to our brain's machinery as we age?

1. Brain responses become less sharply localized. Why is that so important? In a young brain, the brain's response to something happening is a sharp, distinct pattern that occurs on a highly localized scale. In an older brain, the responses are not as sharp, and may be found in multiple brain locations. These changes have usually been interpreted as having compensatory values for the older individual. Perhaps so. It is certainly reasonable to believe that these plastic changes are the result of natural processes in the brain that contribute to its maintaining some level of control. But this slow dissolution of sharply localized activities also reflects a general degradation of the locational specificity of responses generated by "what's happening in the world."

2. Actions within higher brain centers—and the traffic between them—slow down. Even when information is delivered at a reasonable speed, they act on it more and more slowly. Neuroscientists believe that this slowing occurs as a natural adjustment to the deterioration of the quality of information that is feeding these higher centers. Even more importantly, once they operate on it (slowly), they send it to other brain regions on slower transmissions lines that are marked by zones in which the insulation is breaking down on the wires. Our highways and super-highways are developing potholes, and are losing their pavement. Traffic is slowing down. Arrival times are less reliably determinable.

3. System coordination deteriorates. The ability of the parallel sub-systems of the brain to precisely coordinate their actions is fading, in large part because of this growing loss of temporal precision, and the quality of the transmission lines that support it. As a consequence, it is more and more difficult to simultaneously keep track of all of the dimensions of "what's happening" (who, what, when, where) from one source, or as it relates to any single action—and it can become almost impossible to sustain simultaneous, parallel activities (walking and chewing gum at the same time).

4. Predictive bias weakens. The brain has a powerful ability to use information from the present moment to help (positively bias) its interpretation of information at the next moment. This faculty is crucial for reliably representing and completing brain processes, and as we age, it begins to deteriorate. This happens for two reasons. First, because the brain is representing information from the present moment less accurately and less completely, the brain's interpretation of "What's that?" or "What's happening?" is subject to more errors. Second, and probably even more importantly, the brain necessarily takes more time to interpret what it is now hearing or seeing or feeling, and it simply cannot deliver information back in time for it to be useful in the normal fast time that would make it as helpful for representing "What's next?" Biasing in language representation normally occurs syllable by syllable. An old brain has just sorted out the identity of syllable number 1 as syllables 2 or 3 go streaming by. We could maintain control, if the world would just slow down as we get older! Alas, not a chance.

5. The 350 million "micro-computers" in your brain are struggling to sustain their integrity, and are losing their clear messaging. When scientists in my laboratory used recording techniques to assay the teamwork of neurons within the cortex's brain cell "teams" in older brains, we found that the nerve cells no longer operated as strongly in concert. Their coordinated actions are a key for delivering a strong, reliable message to all of the places to which they ship out their "votes" about "what's happening." The degree of local coordination of nerve cell activity has been shown to be directly related to how well the things from the world that have engaged it will be remembered. In the older brain, poor teamwork predominates, and that translates into poor recording and remembering. What's more, as the cooperative activities in these nerve cell teams weaken, their boundaries become more and more indistinct, and an increasing number of neurons struggle to remember which team they actually belong to.

Because of this loss of temporal (i.e., time-related) coordination and the degradation of effective teamwork in the brain's hundreds of millions of "micro-computers," the brain can no longer deliver information in strong forms to effectively engage our highest brain processes. As a result, those higher areas of brain function slowly disconnect from the more coordinated lower levels of the brain's processing machinery. As I discussed in chapter 11, as these systems came online in childhood, these lower levels slowly grew their coordinative powers, and more strongly influenced successively higher levels to enable them to reach their full operational powers. Now, in older age, coordinative powers are weakening at every level in our great brain systems, and the highest levels of functioning—those that are the last to "mature" in younger life—are the first levels to be seriously compromised. The progressive disconnection of the highest processing levels of our brains can be the source of real danger for an owner of an older brain, because these first areas to go offline are exactly the zones where we see the first signs of the development of Alzheimer's disease.

6. Attention resources are taxed. In the fuzzy, noisy older brain, it can be increasingly difficult for the brain to selectively control its attention to favor any one set of inputs or actions over another. Because the neurological representations of "What's happening?" are less salient and more subject to error, it is hardly surprising that we have a little more difficulty generating and sustaining a sharp focus of attention. A strong and clear representation of an attention target is also crucial for the brain's suppression of all of those million and one distractors—things you are not supposed to attend to. This second, "inhibitory" aspect of attention control appears to be especially debilitated in older brains. The older brain usually still does a reasonably good job at turning on and focusing its spotlight of attention (although it is usually a little dimmer and less sharply in focus than before). Unfortunately, that spotlight of attention is more likely drowned out by all of those other lights—the interfering distractors that confuse the attention's spotlight.

7. The brain processes that support new learning, vivify memories, and sustain alertness and positivity are slowly dying. That "modulatory control" machinery contributes critically to your ability to focus and to sustain your attention, to your positive good spirits, and to your self-confidence. Its actions control the "Ready" and "Save it" switches that govern learning. Very significant nerve cell death can occur in these brain regions in older brains.

Even before dying, nerve cells in this machinery simply produce and release much lower quantities of the essential chemicals, those "modulatory neurotransmitters" that enable brain change in learning and memory. When an older person is diagnosed with Mild Cognitive Impairment—again, a level of memory and related cognitive loss that endangers a person's independence—their brain is producing so little of the neurotransmitter acetylcholine that it is almost impossible to measure the levels. Older individuals also produce less dopamine than younger people. Age differences arise partly from dopamine nerve cell death, and from changes in the power of dopamine-induced effects in the brain. You might remember that dopamine is a key neurotransmitter that conveys the "save it" message in learning, and that contributes greatly to our feelings of confidence, pleasure, and success. With dopamine decline, there is usually a parallel reduction in the production of a third neurotransmitter, noradrenaline, which contributes to learning and is a fount of our brightness and attentive alertness.

Many scientists believe that a catastrophic decline in the production and release of acetylcholine presages the onset of Alzheimer's disease. The decline of dopamine production and release—which is usually closely paralleled by a decline in noradrenaline—is a signature marker of the onset of Parkinson's disease. Alzheimer's and Parkinson's symptoms are often recorded together, as these systems can slowly die off together, as they do towards the end of many people's lives. The reduction of noradrenaline release presages a slow lapse into clinical depression.

The brain activities that control the release of these crucial modulatory neurotransmitters not only enable brain-controlled learning, but also play important roles in regulating sleep. If you struggle to sleep through the night, or you struggle to awaken with your former liveliness, your brain's machinery controlling these key chemicals may be in need of repair.

8. Blood supply is at an all-time low. Just as the brain circuitry is simplifying and pruning, so, too, is its local blood supply. Blood delivers essential oxygen, sugars, and other nutrients that sustain the brain and support its activities. Good blood perfusion helps sustain your immune system in good stead. The ability of the vasculature to respond rapidly to demands for fuel just where and when it is needed is also slowly declining in older age. And the machinery within nerve cells that uses this oxygen and food to sustain their health is weakening decade by decade.

9. The immune system is compromised—and that has major consequences for the brain. The immune system has the job of cleaning up the mess that comes from neuron simplification and loss. When the brain loses its capacity to clear out the physical debris generated by cell death and connectional pruning, bad things can happen. I'll talk more about that in chapter 26.

At first reading, you might be thinking that all these changes present a bleak picture for the later decades of your life. But here is the good news: almost every negative change that we've just described can be driven back in a positive direction by specific forms of brain engagement and brain exercise. You might have let things slip a little— but you are endowed with that great gift of plasticity, and as long as you're still alive and kicking, you can still use it to drive your brain back correctively. Remember that brain plasticity does not just affect the functionality of your brain; it also provides a powerful basis for <u>physically</u> growing or restoring your brain. You can fight against these negative changes to push your brain in the direction of restored competencies, to recover at least a little—and perhaps a lot—from nearly all of markers of the average older—or otherwise neurologically challenged—life and brain.

For further explanations and extensive references and citations related to the information in this chapter, please visit
www.soft-wired.com/ref/ch22

23

WHY IS MY BRAIN SLOWLY LOSING IT?

Root Causes of Changes in the Aging Brain

It's common wisdom that the brain is going the way of all flesh—like all machines of an older age, straight to perdition. To quote a few authorities on the subject:

"When I was younger, I could remember anything, whether it had happened, or not; but my faculties are decaying now and soon it shall be so I cannot remember anything but the things that never happened."
-Mark Twain

"These days I can walk into a room full of people and the only name I can remember is Alzheimer." - Anonymous

"Life is a moderately good play with a badly written third act." -Truman Capote

Yet there have always been exceptional individuals who have developed strategies for making the best of life's last chapters.

"The idea is to die young as late as possible." -Ashley Montague

"How old would you be if you didn't know how old you were?" -Satchel Paige

"Anyone who stops learning is old, whether at twenty or eighty." -Henry Ford

My personal favorite in this vein is something the great cellist Pablo

Casals is reported to have said. When he was in his eighties, Casals, whose professional career spanned more than seven decades, was asked, "Why do you keep practicing?" His response? "Because I can get better."

Most scientists who study the neuroscience of aging are in the business of documenting, in exquisite detail, one or more of the hundreds of ways that the brain physically, chemically, and functionally deteriorates—and how those changes correlate with the decline in our behavioral abilities as we age. I've done quite a bit of that myself in our narrative up to this chapter. The predominantly held conclusion has been that the brain, like every other heavily used machine, just wears out. Its cogs are worn, its gears are loose, its tires are bald, rust and grit accumulates, and its shock absorbers are completely shot. Its slow decline ultimately results in changes that may lead up to a final, more rapid, catastrophic collapse that is expressed in many people as Alzheimer's or another of the many neurodegenerative diseases out there.

Of course, from my perspective, there is a lot more to this story, and it involves brain plasticity. The development of a healthy and highly functional brain is a consequence of how a person has used it. The brain is plastic—it can change—and what's more, every person has the capacity to control positive plastic brain change in their own brain.

So why does the brain deteriorate? There are several important answers to this crucial question. You should not be at all surprised to hear that we've already described part of the answer: Older brains are, to some extent, just wearing out. As I discussed in the previous chapter, the ability of the brain to sustain itself physically is affected by the slow deterioration of its vascular supply, by changes in the cellular energy, nutrient sources, and immune system assets that are crucial for its health, and by the slow loss of nerve cells and other elemental brain structures. While these changes can be substantially lessened by using the brain effectively, no known behavioral experience-generated transformation can totally reverse them.

Other age-related changes are occurring in the heart and lungs, endocrine glands, and other internal organs that support general brain health. This general wear-and-tear model also applies to eyes, ears, and other sensory organs. As the quality of information from vision, hearing, skin senses, taste, and smell slowly degrade, the plastic brain makes natural adjustments that can actually amplify the degradation in the quality of information that you extract from them. In other words, when you are seeing, feeling, listening, sensing, smelling, or tasting, your brain is getting less information from the senses and the even-poorer use that it makes of that more-limited information may add to (not help overcome) this problem.

Finally, muscles, joints, bones, and the related tissues that control all of your actions are also physically deteriorating. Although we all know that strength, balance, and flexibility are dependent upon a regime of regular exercise, most people admittedly don't exercise with enough consistency or frequency. But even if a person's exercise regime is undertaken seriously and in the most appropriate forms, some inexorable decline would obviously still be recorded. It is an undeniable fact of life that we have not been constructed to live forever. The goal is to improve the operational capabilities of your brain—especially if it is older or damaged—as much as possible.

You might already intuitively understand a second reason why the abilities that defined a person at their peak will slowly deteriorate. For the average older citizen in the modern, industrialized world, the brain suffers from disuse. As we age, we may think we should "take it easy"— finally be able to relax, no longer having to put up with the stresses and struggle of the daily grind. Enjoying the golden years is easier. For older people who continue to work, they may feel they have totally mastered their job, and are now running in automatic pilot mode.

The truth is that most older people have long-since entered into— and at least begun to accept the "fact" of—the decline of their brains and bodies. I mentioned earlier that for most people, the apogee occurs sometime during the third or fourth decade of life, and for some (especially women) perhaps in the fifth decade. During this first several decades long epoch of brain growth, almost every day involves learning skills or continuing mastery of abilities, in a life that is almost continually challenging. Experience-driven brain plasticity dominates the days of childhood, youth, young adulthood, and early professional development. A million and one challenges in skill acquisition and the development of more refined abilities account for an explosive growth of—and a high level of precision for—our neurological faculties. Whether a person peaks at age 25, 35, or later, they possess tremendous capacities as an accurate receiver and facile user of all kinds of information—and the brain is able to handle multiple, fast inputs with speed and clarity. During this period, we remember most of what we hear and see. We construct particularly rich associations of information from all the senses at high speed and with great efficiency and accuracy.

However, as our skills reach a level of mastery one by one, for an increasingly greater part of each day, we come to be mere users of what eventually grow to be automatized abilities. Once any particular skill or ability advances to what we judge to be an acceptable performance level, we quite naturally begin applying it on automatic pilot, with little effort expended toward improving it further. In short, many older adults spend most of every day as living, breathing automatons!

On some level, our brains are actually designed to do exactly this. As I noted earlier, the great philosopher and psychologist William James beautifully described this process more than 120 years ago. We humans, he argued, develop reliable behaviors (he called them "habits") across our earlier life. Once established, we employ these habits effortlessly, without thinking about them. That frees up our "powers of mind" for the more complex aspects of behavior that account for our intellectual flowering and our more sophisticated abilities. The deeper and broader our platform of earlier-constructed "habits," the more resources we have to draw from, to operate on the highest planes of active thought and action.

To better understand how you evolve from a learner to an automaton, consider the skill of typing. Many people learn to type at some point in their youth. As a kid or teenager, they may work very hard—as I did in high school—to achieve high proficiency. However, at some point, typing is judged by the brain to be acceptably "mastered," and most people stop working at getting better at it. They become users—no longer attentive learners—of skills that reach a level of practical utility.

I won the Typing Award as a high school sophomore. Would I still win the Typing Award at this point in my life? If I took a typing test today, would I type as many words a minute with as few errors as when I was at my peak? For this to be true of people over about 50, the typist in question must either be exceptional or delusional. The older typist is highly likely to have gone through the same progression from learning to mastery to using, in most of the crucial skills and abilities that define your effectiveness. The lack of practicing in the crucial domains of listening, seeing, movement control, and even complex thinking, is the grease on the long, slow slide to cognitive decline.

It doesn't have to be that way. Pablo Casals really paid attention, and was actively striving to improve, as he practiced that cello with intensity every day of his professional life, almost to the day that he died. Because he sustained a refined status and continued striving to improve, he never lost his abilities as an accurate listener. He never lost the ability to control incredibly precise finger and hand movements in unbelievably complex and rapid sequences. He never lost the ability to profoundly express his emotions through the movements of his hands and body, with impeccable precision and power, or to confirm their meaning through his active listening. He never lost his abilities to translate the orthography of the musical score or the cueing of a conductor into his exquisitely controlled hand movements, confirmed by extraordinarily refined listening skills.

Pablo Casals is an outlier; most people don't practice a musical instrument or other complex, multi-sensory skill late in life. But without those appropriate forms of practice, you simply cannot maintain the high-fidelity representations of the information that the brain is receiving and interpreting on the way to controlling its actions in movement or thought.

I noted earlier that my earlier research has shown that a researcher can accelerate the aging process by adding "noise" into an animal's environment, or by "injecting it" directly into the processes in the brain. The "noise" affects a negative change in the sharpness of the brain activities that represent "what's happening." When the brain struggles to resolve what it senses, hears, sees or feels because its machinery is plagued by background "chatter," the processes of memory, thought, and movement control deteriorate. Fuzzy brains must take longer to get the answer right. When that "noise" is present, the signals from the eyes, ears, and body are fuzzy and indistinct, so more concentration and more time is required to determine "what's happening." Imagine that you are out at dusk trying to make out what's in front of the trees across the meadow. In the gloaming, it will take quite a bit more time to resolve just what it is. "A deer," you tell yourself, as it begins to move off into the wood. An older brain may have the same problem even on a sunny day: The signal—in this case, the poorly resolved vision of the deer—is weaker. To make matters worse, the background—the dark meadow and sky—is proportionally stronger than in a younger brain. Under these poorer conditions of light and contrast, getting the answer right takes longer. If an older brain takes longer to get answers right, it plastically strengthens just those connections that favor it taking longer, at every step that it takes in making a decision. The brain's stability requires that it consistently get the answer right. The brain's slowing down represents a completely natural adjustment by a brain that is struggling to do the best it can to retain control under increasingly difficult (internally "noisy") circumstances.

A degradation in representational sharpness and speed occurs for any skill that is left unpracticed. Reflect, for a minute, on what would have happened to Mr. Casals's cello playing if he had stopped practicing his instrument at age 35. By age 40 or 45, he would probably no longer be a world-class cellist, and a propensity for small errors and less precise instrumental control would be obvious. At age 50 or 55, he would no longer remember much of his former repertoire, or be able to play the most difficult (especially the speed-challenged) musical passages that he had played earlier with high proficiency. By age 60 or 65, he may still be a pretty good cellist, but no longer a significant, much less a great, artist at all. He would not be able to play the more challenging music in his earlier repertoire to his satisfaction. He would not be able to learn new pieces as easily. Thank goodness Mr. Casals knew better!

Unfortunately, most people don't know better. Most people don't make dedicated efforts to stay sharp in the crucial skills that would assure the maintenance of a good memory, mental agility, or the facile control of speech and movement abilities.

One question this brings up is: How about riding a bike? People never forget how to ride a bike! If you're still riding regularly, you're practicing on some level under appropriately high-stakes conditions. But if you're not, hop on one, and see how you do. If you think that you can still ride with all of the control and flair that you had as a kid, you might be in for a surprise. You might be able to recover most of your old skills, but depending on your age and how long it's been since you've been on a bicycle, it may take a very considerable amount of new practice.

Because of its under-utilization, the learning-control machinery in older brains is slowly dying. Most older individuals believe that a key to brain health is "staying active." Instinctively for them, that means active reading, playing cards or other familiar games, spending a generous amount of time on the golf course or tennis court, taking an adult education course, meeting socially with their friends and acquaintances, traveling, and so forth. These things will help a person "stay active"—but because they tend to be non-demanding and have already been developed to an acceptable level of proficiency, they probably aren't helping your brain health very much at all. If you're comfortable with a skill, and are no longer pushing yourself, then all the "staying active" in the world isn't going to provide much benefit for your brain. The brain is a learning machine. It is begging for new learning. Especially past roughly the midpoint of life, most adult individuals are feeding it nothing but the same old stuff. And to the extent to which they view themselves as "still learning," they are confusing "content acquisition" with learning—which, in the context of the brain, are not the same.

The brain machinery that produces those chemicals that inform that brain that it's "ready" for change, that increases the repertoire of those things that, through learning, the brain can change to, and that turns the change switches to "ON" and "Save it" is sadly neglected in most old brains. This same machinery controls how bright and awake you feel every morning and throughout every day; how ready you are to drink in your physical world; how on-the-ball you are in the world of thought; the levels of warm feelings and joy that should be a part of almost every really good day; and how readily you can turn out the conscious lights at night, as you put your brain to sleep. As a function of how active they are, these brain regions release chemicals known as "growth factors" that increase their energy stores, increase their production of those crucial modulatory neurotransmitters, and increase the overall vitality of this important machinery. In most old brains, because they just don't get anything close to the exercise they need, all of those stimulating and learning-enabling chemical effects become weaker and weaker. And because the production and release of these growth factors are not being stimulated, the very health and survival of this critical machinery is put at risk. Its tailspin—added to by the related losses of memory—is a main contributor to an older individual's lowered learning rates, and to their slow loss of bright spirits, self-confidence, and a full measure of happiness in their older lives.

Older brains operate more often at a level of abstraction, because they are struggling to reconstruct the details of what they see or hear or feel. There is another facet of these age-related changes that few older individuals are aware of. In the natural progression of how a person learns to operate with mental agility, the brain progresses from struggling with the details of seeing, hearing, feeling, tasting, smelling, and acting to the ultimate creation of beautiful methods of abstraction that no longer require all those details. Learning to read provides a good example of how we learn to take in information at multiple levels of abstraction. Reading training begins with learning the alphabet and learning to understand and vocally produce sounds, which is followed by mastering the details of which sounds go with which letters. That leads to the ability to read syllable by syllable, then word by word, then phrase by phrase. In the beginning, reading is all about the details. But once you master the skill, you primarily operate at a higher level of abstraction, swallowing large chunks of text at each bite. A first grader can only read by focusing on details; a highly proficient reader isn't even aware of those details.

If a person is reading on the abstract level and starts to zone out, they "down-check" by rapidly moving the eyes and shifting to focus attention back in the text. The details can get the reader back on track—

and the active attention to details generates more precise and more coordinated neurological responses that, for the brain, are far more easily recordable as memories.

Scientists have elegantly demonstrated this capacity for representing the same information at multiple levels of detail and abstraction in the domain of speech and listening. They can dramatically simplify the sounds that represent aural speech in an adult so that all that individual hears are the fluctuations across broad bands of energy, with almost no resolution of the different pitches of sound in the speech. Amazingly, though devoid of details, most adults can still fully understand this kind of speech. The brain has gradually learned to understand it because this abstracted form of information is also always present, and is highly correlated with the detail-filled forms of speech that were first learned as a child. However, even though you may understand it, you can only very poorly remember it. Your remembering specifically requires that you encode a few more of those details! Moreover, add just a little "noise" to this abstracted acoustic form of speech and you can no longer understand or remember it at all.

Scientists have also created a form of acoustic speech that specifically represents only those little detailed moments of change. This detail-rich form of speech is entirely different from the abstracted form that cannot be remembered. When a person hears it, the brain generates sharp, strong, synchronized responses in the brain areas dedicated to speech-reception. And it can be remembered. With this detail-rich form of speech, nerve cells are representing those sharp details with powerful teamwork, and any interfering noise has to be very loud before it can disrupt your ability to understand or remember it.

Older brains are very reliant on the "quick and dirty" strategies that the brain has developed in things like speech reception, reading, seeing, and driving, among hundreds of others. This is in large part because its abilities to accurately record the details of "What's happening?" are slowly declining. Because we still understand "What's happening?" on this secondary, less-precise, abstracted level of analysis, the decline of our more fundamental faculties is hidden to us. We think we're doing just fine. Because we can be ignorant about the decline of these crucial, elementary signal resolution abilities, we have no intuition about what's really wrong with our brains, and we neglect the kinds of detail-level behavioral practice and effort that is required to sustain or rejuvenate them.

An understanding of how what you are not doing can contribute to slowly deteriorating neurological competencies leads to an understanding of some of the very obvious and potentially powerful positive things that could contribute to better brain health. I'll get more specific about how you might initiate a positive course correction in life in part five.

For further explanations and extensive references and citations related to the information in this chapter, please visit www.soft-wired.com/ref/ch23

24

PUSHING HARDER ON THAT ACCELERATOR,
IN REVERSE

How "Negative Learning" Can Speed Up Brain Decline

Now that you have an inkling of some of the things that people don't do, and that can potentially contribute to slow changes that can frustrate an older life, it is important to consider some of those things that you are doing that commonly accelerate the rates of backward change. It does not occur to most individuals that in their attempts to sustain an acceptable level of control, they may have adopted destructive, learned compensatory strategies.

"How on earth," you may be asking, "can compensation be destructive?" Let me illustrate how negative learning contributes in a major way to your losses as you grow older, by describing four simple examples.

You fall, and hurt yourself. You realize, in falling, that you could not have stopped your fall. You stepped on that small pebble just as you were jostled, and down you went with nothing to impede your collapse. An unpreventable fall or two or five can be frightening enough to encourage a person to take steps to prevent any recurrence. "I must make adjustments to walk more safely," you tell yourself. You instinctively turn your head down more often, to watch your feet; you regularize the control of your stepping; you flex your hips and knees, and separate your feet just a little to stabilize your stance, believing that, like the dancing boxer, it might help your make a faster, corrective response.

You haven't really been learning about postural control or walking since you were a toddler. Now, you're learning again. Unfortunately, nearly all of these instinctive adjustments intuitively designed to increase your safety result in negative plastic changes that bring you closer to a wheelchair, rather than further away. By making them, you have taken significant steps toward losing your mobility. To understand how, let's review their neurological consequences one by one.

First, by turning your head down to watch your feet and the ground in front on you, you've changed the functional axis of the balance organ in your head to a less effective mode. You've been using the same balance mode—the head upright with eyes looking straight ahead to take in the world in front of you as you walk—for your entire life, and it's worked pretty well up to this point. Now, after decades of successful walking marred by a few falls, you decide to convert to a new, head-down mode.

By turning your head down, you sacrifice information about things moving rapidly in front of you that you must see and react to if you are to avoid being knocked to the ground. And when you begin to fall, for any reason, the information that you're receiving from your eyes now comes from the very-fast-moving ground in near vision, rather than from the slower changes in the visual scene that would occur if you were equivalently bumped while looking forward. The movement of the ground before your eyes when you're looking down now occurs too rapidly for you to use to make corrective postural adjustments. If someone bumps into you while you're looking down, that fast-moving scene is very likely to carry you (in a sense) right down to the ground.

By separating your feet and flexing your knees and hips to achieve what you instinctively believe to be a safer, more active stance, you've substantially increased the energy requirements for standing and walking. In standing postures, we humans most efficiently support weight by translating it straight down through the long bones in our legs. When we flex the knees or hips as we stand or walk, our weight has to be continuously supported by always-active muscles on both the fronts and backs of the legs and hips. It doesn't take very long for us to tire. And once again, the starting point for the initiation of all of our corrective movements from this new stance has been altered, from one that you've been using your whole life to a new one that is supported by far less prior experience.

Finally, you've made the especially destructive adjustment of regularizing your stepping so that your walking is more consistent and your feet are always close to the ground (and therefore never in danger of making a reckless move). Now, you don't walk, you shuffle. It's downright dangerous, because through that learning, you are adopting a new form of stereotypic walking that is no longer effective, neurologically, for handling a surprise. It is surprises that lead to most falls. In childhood, we develop a marvelous capacity to walk and run and gambol across the landscape under almost any condition. The pebble underfoot, the slippery patch, the unexpected bump on the shoulder by our playmate presents no problems for us after a few years of practice, because rapid adjustments to almost any variation in the conditions of walking come to be safely within our behavioral repertoire. Now, the compensating older adult is voluntarily sacrificing all of these dimensions of free-form, anything-goes walking for what they believe to be a reliably safer, more regularized form of walking.

None of these things make you safer. By these steps, you have employed your brain to degrade and impoverish (not sustain and enrich) a crucial set of motor skills that your very independence relies on. With rare exception, once you start down this path, if you live long enough, you can look forward to the day when standing is torture and walking is a memory.

Let's consider a second, simple example. It seems like you can't quite follow along with the radio or TV as well, so you turn up the volume. Accuracy in reception fades a little more. Up goes the loudness. Accuracy fades again. Let's make it still louder. Pretty soon you've got the loudness control turned up to the max. Pretty soon a stranger entering your home is in danger of incurring a hearing loss because of your TV volume. Indeed, so are you.

You probably learned at a young age that if the station isn't tuned in correctly, you turn the tuner knob to fix it—not the volume knob. For your brain, you're turning the wrong knob!

By turning up the volume on your TV, you are specializing your brain for louder and louder talkers. This would be a great strategy—if only you could get all the people out in the world to talk a lot louder, too! As it is, it's a contributor to further communicative loss and incompetence. If your audiologist tells you that you really do have low-volume hearing, get a hearing aid so that every source of speech is equivalently louder. Whatever else you do, think twice about using the volume knob liberally on your television set or radio as a substitute for one.

My third example comes from another consequence of sitting all of those hours in front of a television, computer, or iPad screen. If you're a typical citizen in the U.S., you spend an astonishing 50+ hours a week consuming media from screens. "How," I can hear you complaining, "could watching screens drive negative changes in my brain? I've been doing that since I was a kid, and it hasn't hurt me yet!"

Perhaps it has not occurred to you that focusing your vision on the things right in front of your nose for most of your hours occupying most of your days is not natural. In our natural environments, our brains were wide open to the world in front of us from ear to ear, because survival was dependent upon our rapid interpretation of and response to unpredicted events occurring anywhere in our immediate environments. That natural progression from incoming information to action is accomplished through circuits that heavily engage the frontal lobes of our brains. Guess which brain areas take the strongest hit in a couch potato?

Just as importantly, we now spend much or most of each day engaged in visual behaviors that constrain our field of view to that smaller box-like zone immediately in front of our eyes. Everything outside that box is a "distraction," deemed by our brain to be unworthy of our attention. Our brain is very good at adjusting its machinery so that all of those distractions away from the narrow view directly in front of us are slowly, enduringly suppressed—and ultimately, unseen.

We also exaggerate this contraction in our visual view when we drive our cars. Moving in a heavy machine at high speed on raceways occupied by other fast-moving machines requires our sharp, focused attention on what is directly in front of our noses. That narrow span in visual attention must be constant; it can be fatal to us if it flags for even a few seconds. This high-stakes control of our vision for those things happening in the "center of gaze" results in the slow degradation of our ability to keep track of a million and one possible things that might occur in our peripheral vision.

Besides the systematic starvation of our frontal lobes, there are three other serious, negative consequences of our developing great powers of sustained attention in the central zone of our field of view.

First, our view of the world slowly contracts. By the age of 60, we've lost about a quarter of the visual perspective in our field of view. By 80, we've lost roughly half. What if I were to say, "From now on, you're going to have to get by with a TV that is 25% smaller—50% smaller— than the one that you have in your house"? My guess is that you might not want to swap your big-screen TV for the smaller model that I'm offering.

Second, we spend an enormous amount of time engaging in activities that are important to us, practicing all that staring at that smaller world confined by the edges of our TV, book or iPad, or by the demands of unflagging attention to the road ahead. As a consequence, our vision is progressively more weakly attracted to any of life's innumerable little visual surprises. Our eyes move less often. When they move, it is often from stare to stare. What if I were to say, "From now on, I'm going to put Botox in the muscles that control your eyes so they can only stare straight ahead"? My guess is that you would not be in favor of that procedure.

Third, our brains are invigorated by our paying attention to the unexpected. Surprises propel our brains and bodies into action. The machinery engaged by all of those little surprises in life controls the brightness of our spirits, and the spring in our steps. What if I were to say, "From now on, I want you to know that most things that used to be interesting that are happening in your world are just really not that important, and I suggest that from now on you just ignore them"? I would guess that you might tell me to cease these foolish suggestions, because you know that those little things that spur your interest and attention are a significant part of the very spark of life.

It is rather paradoxical that one major impact of these self-induced neurological changes is its impact on our safe operation of motor vehicles. When we're hustling around in those large, fast-moving machines, it is a very good idea to be able to detect and respond to those very infrequent but potentially deadly things that can be coming at you out of your visual periphery. The accidents of young drivers largely involve their still-limited ability to control their unflagging attention on what is happening in their visual world out in front of them. They most often run into that car they are following, or into that tree or fireplug or light post along the margins of the road. Older individuals, with all of that TV & computer screen and automobile driving practice, are absolute masters of the straight-ahead. Older people are far more likely to have accidents at an intersection or in the parking lot, where we hit (or are hit by) other vehicles that come at us from the reaches of that limited peripheral vision.

For my fourth example, I want to tell you a little story about my childhood. My mother was a piano teacher, but of her six children, I was her most notable failure as a music student. I was too fractious, and always pined to be outdoors at play whenever she tried to sit me down on the piano bench. After about a year of poor cooperation on my part, she gave up, to both my and her considerable relief. As a little older boy of 10, I saw that other children were signing up for the band, and decided, a little jealously, that I didn't want to miss out on the fun. After lots of pleading necessitated by my earlier musical failure, my parents bought me a shiny new alto saxophone.

At that age, I was determined to succeed, in part because of that sense of competition that can arise as a pretty strong force in a child.

But initially, like most beginning musicians, it was a struggle. I was all squawks and squeaks and clumsy fingers. I began to tell myself, in a growing panic over a period of several weeks, "You can't, you can't, you CAN'T do this." Almost every child has found their way past a sustained personal panic attack like this, on the path to discovering that they actually can shoot a basketball, sing well enough to get a part in the musical, or solve a tricky algebra problem.

At some point in my panic, over a narrow span of time, I shifted each of those "You can'ts" to a "You can." And my brain could. Within a few days I could play the saxophone. Within a year, I played Rimsky-Korsakov's "Flight of the Bumblebee" on the local radio station.

It is tremendously destructive for you to tell yourself that "you can't," if, in fact, succeeding just requires a little more serious effort and practice on your part. Your brain registers all of that negative messaging. When you tell your brain "I can't" just a little too often, it becomes a self-fulfilling prophecy.

It is distressing to have to hear about all of those things that older folks have lost their ability and will to do. "I don't ...," "I can't ...," "I wouldn't still ...," "I no longer ...," "I used to be good at ...," " I don't think I ...," "It's now too hard ...," "I really don't do that anymore ...," "Not for ten years ...," "I'm too old for that ...," "I used to like ...," "I used to be good at ...," and so forth. Sometimes your brain should be told that it can't. Some people drive too long. Some take risks at an older age when they should arguably be more careful. But for every mistake on the high-risk side, there are a dozen or two questionable decisions for most older folks on the low-risk side. Giving up those harder things in life just because you no longer choose to be challenged by life in any significant way can be self-destructive to independence and health.

We have already discussed another aspect of "negative learning," when I explained that one destructive feature of most modern lives in a greatly reduced schedule of new learning. Alas, much of the sparse new learning that older individuals actually engage in falls on the "negative learning" side of the ledger, usually consciously designed to compensate for our progressive decline—but at least as often as not having almost exactly the opposite impact on our brains. For many older individuals, those compensatory behaviors are designed to achieve what we perceive to be a great personal goal: An easier, more trouble-free older life. Compensation and gradual mental simplification in our older lives are consciously and subconsciously designed to meet our goal of achieving a life of ease.

Of course modern citizens are also supported in their life of ease by technological advances that are designed to help you sleepwalk through life. The average older individual passively observes for many hours every day, rather than actively doing something. There's not much real new skill learning there. If a task is too challenging for us—actually challenging our brain to really focus our attention on it much less to learn and really improve at it—people generally look for a technological answer that will quickly remove that challenge. Or, they'll just skip it altogether.

To cite one of a thousand examples, it is useful to practice your navigation in the world by recording the complex serial information that flows by in the visual scenery as you roll (walk, run, crawl, drive) across the landscape. The health of your hippocampus and the library of your long-term memories need this kind of exercise. Of course modern humans don't really need a brain to navigate across the landscape—we can rely on a Global Positioning Satellite (GPS) device.

A British research team looked at the brains of adults who had just obtained their licenses to drive a London taxicab, which requires a four-year driving apprenticeship, and the demonstration that you have a detailed mental mastery of the streets of Greater London. Not surprisingly, when the researchers looked at each cab driver's hippocampus—that brain region that helps us construct our world in the dimensions of space and time, and that crucially supports our long-term memory—they found it was substantially larger and much more strongly activated than that of the average London citizen. Scientists have exercised rodents in an analogous way, by providing them with new navigational challenges every day, as a prerequisite for their receiving their daily grub. Their hippocampus grows apace! Big hippocampi are a very good idea, memory-wise. And I'll talk about in chapters 28 and 29, re-growing your hippocampus is one of several very important antidotes for the dreaded Alzheimer's disease.

More recently, scientists have evaluated the consequences of an extended period of GPS use. Guess what happens in your brain? These same abilities that support your reconstructions of that world out there and that sustain your wonderful memories about it take a big hit. To the extent that you adopt these kinds of technologies to hold your hand throughout life, and to the extent that you take your brain out of action so that nothing in your day is cognitively challenging, you have committed yourself to an even faster rate of decline in your movement, memory, and thinking.

You might be interested to know that there is a very dedicated technological research and commercial effort directed toward developing strategies to monitor how older folks are doing in their daily lives, then applying technological solutions to assist them to assure that they get it done right. These technology futurists foresee wiring you and your environment with a variety of detectors that assay "What's happening?" that will monitor your responses, drive your car for you, and in many other ways help make sure that you respond to the world appropriately. They have already developed a wonderful strategy to solve just about every problem for you by an answer-lookup strategy that we all use almost every day—Internet search. They dream of a wearable brain-like computer that you can carry around on your shoulder or in special glasses that can see and hear everything you do, remember everything, and be less emotional and always right. Perhaps we can look forward to an older life that just doesn't even require that we have brains at all!

Only teasing.

For further explanations and extensive references and citations related to the information in this chapter, please visit
www.soft-wired.com/ref/ch24

25

FRANKLY, THIS SUBJECT IS DEPRESSING ME

How Other Medical Issues Contribute to and Accelerate Decline

I have a friend, Colleen, who was bitten by a tick at her country home in Pennsylvania about 13 years ago, when she was about 50 years of age. That tick injected her bloodstream with Borrelia spirochetes. She was aggressively treated for the infection with antibiotics, but that treatment did not prevent the Lyme disease infection from spreading into her brain. Eleven years later, she is at home, close to the end, with an advanced form of Alzheimer's disease (AD).

A brain infection like hers is just one of many factors that can increase the risk of an earlier onset of dementia, or of Alzheimer's itself. Chronic high blood pressure, alcoholism, multiple sclerosis, uncorrected hypothyroidism, obesity, serious hearing impairment, HIV/AIDS, uncorrected folic acid deficiency, PTSD, a career as a professional or amateur boxer, chronic depression, chronic tinnitus, encephalitis, untreated vitamin B12 deficiency, heavy metal exposure, open heart surgery, playing football, malarial brain infections, OCD, chronic drug addiction, playing hockey, blindness, epilepsy, Downs Syndrome, loss of mobility, prion disease, bipolar disorder, traumatic brain injury, ADHD, Parkinson's, stroke, schizophrenia, PCB exposure, Huntington's, chronic anxiety, living in Tennessee or Washington, meningitis, chemotherapy, and a growing list of combinations of widely prescribed drugs all add to your risk. While we have problems explaining why there is a higher incidence of AD in the states of Washington or Tennessee than in other states, almost all of the other factors on this list generate changes in your brain that increase the "noisiness" of its operations in a way that can be expected to contribute to the acceleration of your cognitive losses with age.

You might be interested to know that there is absolutely no doubt that our doctors and scientists have not yet come close to completing this list. The real list almost certainly has several hundred (and probably thousands of) entries.

That's pretty surprising, if you go to someone that you think might be an authority on this matter, like the American Alzheimer's Association. They list two main risk factors: age and genetics. As for getting older, sure, younger is better, for not having AD. As for genetics, they are definitely a contributor—but you have to consider all of the other things that could have contributed, like how well you have taken care of your body and brain, as well as the long list of other possible complications listed above. Cumulatively, unless you are genetically at high risk for early AD onset, these many neurological vicissitudes that befall the majority of people, combined with how we have (or have not) led a brain-healthy life, can more than counterbalance any genetic advantages or disadvantages that you may have.

By the time that we pass age 70, the majority of us have had something happen to us in our medical lifetimes that have bounced us out of the "normal aging" category. Fortunately, with a few exceptions, the principles of the processes by which the brain can plastically improve its faculties are only rarely incapacitated by these other brain insults. Now that you have an inkling of some of the things that people don't do, and that can potentially contribute to slow changes that can frustrate an older life, it is important to consider some of those things that you are doing that commonly accelerate the rates of backward change. It does not occur to most individuals that in their attempts to sustain an acceptable level of control, they may have adopted destructive, learned compensatory strategies.

For further explanations and extensive references and citations related to the information in this chapter, please visit www.soft-wired.com/ref/ch25

26

ALZHEIMER'S DISEASE

How Changes in the Brain Can Lead to a Catastrophic Collapse

When I first established a company dedicated to helping older people sustain their brain fitness[8], we conducted a marketing survey to try to determine the perspective of older American citizens about their normal aging. Fifty hour-long interviews of 60- to 80-year-olds were recorded. Although no questions on this survey were directly related to Alzheimer's disease, an astonishing 49 of 50 individuals raised the issue on their own. As the 50th interviewee was leaving the room, he turned back to the interviewer, raised his right hand with index finger elevated to make a final point about "one last thing that [he] should have mentioned ...." whereupon he described his palpable fear of a later decline into Alzheimer's disease.

We are almost all afraid of this bad ending. We've all seen it happen often enough for that fear to float across the shadows in our mind every so often—and as our own impairments grow in ways that potentially threaten our independent living, those worries can sometimes move from a misty background to confront us directly.

Scientists call Alzheimer's a "neurodegenerative disease." That "neurodegeneration" is expressed by two main elements: the growth of poisonous crystals of a substance called "beta-amyloid" in the brain, and nerve cells that are functionally disabled by the abnormal proliferation of those snake-like structural elements we've talked about earlier called "microtubules." Any brain cell in the neighborhood of an amyloid crystal is rendered inoperative; brain cells filled with "microtubule tangles" are also functionally inoperative. These dead zones of toxic crystals, tangle-

---

[8] Posit Science Corporation, headquartered in San Francisco

filled nerve cells, and poisonous soluble amyloid—from which the amyloid crystals are formed that is just floating around causing havoc in the brain—grow like the measles, commonly over a period of a decade or two. In the early stages of the pathology, the distribution of these poison spots is sparse, and the areas of the brain that are affected are limited; the "contagion" gradually spreads more widely across the forebrain, with the dead "spots" slowly increasing in their numbers and sizes.

Why, exactly, do these changes result in catastrophic changes in memory and other cognitive abilities that rear their ugly head as senile dementia in an older individual? I earlier described how your brain had to work so hard to organize its activities in detail, to create its representations of what's happening in your world. Those 350 million "microcomputers" have to operate with a high level of teamwork to create accurate and reliable reconstructions of "what's happening." They have to be able to receive and analyze those details at high speed. Strong and accurate forms of the representation of "what's happening" have effectively fed that information to the brain regions that account for our highest powers. But as the years pass, the older brain has been slowly losing its speed and precision because of a lack of adequate inputs. Other genetic and environmental factors have also contributed to an increase in the "fuzziness" of nerve cell responses as the brain struggles to represent and record the details of what's happening. Now, with emergent AD pathology, someone has begun to throw little grenades (amyloid crystals and patches of high levels of soluble amyloid; patches of "tangle"-filled neurons) into the machinery. Everywhere one on those grenades goes off, the machinery of the brain has an even more difficult task of sustaining nerve cell coordination (teamwork). In the face of all of this rapidly growing internal chatter, the affected regions of the brain struggle mightily—and ultimately catastrophically fail—to make sense of "what's happening."

When those grenades first begin to fall, you're almost certainly still doing fine, and quite a few of those little bombs have to be thrown, as a rule, before things have deteriorated to the extent that your doctor pins the AD label on you. If you're over 60 or so, it's pretty likely that Mother Nature has already begun tossing a few of these little grenades your way.

What does brain plasticity science tell us about the true causes of the disease itself? Why does the pathology begin to grow in your brain? In the last chapter, I mentioned some of the things that have been shown to contribute to an earlier onset of AD. I noted that at least most of these factors can be expected to add "noise"—internal chatter—to our neurological processes, degrade the refined processing of information in our brain, and accelerate or amplify the negative changes that are recorded in normal aging. By that perspective, these additional risk factors just carry us to the edge of the AD cliff faster than would otherwise be the case. Does that mean that Alzheimer's disease is a predictable, natural consequence of growing older?

Yes—but you have it within your power to do a lot that can be expected to greatly slow down these destructive natural processes.

What actually sets the brain off on the path to filling itself with poison crystals and tangled-snake-filled nerve cells? Consider a brain that is slowly deteriorating due to normal aging. Those negative changes would have been more rapid if the brain had been neglected in all of the ways that I earlier described. As I noted in the last chapter, the majority of us also carry significant additional burdens that have contributed to that decline, stemming from our earlier medical histories. As the brain slowly loses control, the first areas to go off-line are those regions of the brain that were the last to go on-line. As you recall, those areas contribute to our most sophisticated ("highest") levels of control. Regions in the frontal lobe (behind your forehead) and ventral temporal lobe (just above your upper jaw) that control our most powerful and most complex operations in prediction, thought, social interaction, and problem solving are among the first to drop out. Brain areas that feed the hippocampus, that crucial machine that hangs our memories on the curtains of place and time, are also among the first areas to become functionally inactive. In earlier life, these areas were continuously engaged when we still retained our peak operational powers in our younger life. Now, they are almost completely disengaged. Without activity, their inactive synaptic connections—and later, the nerve cells themselves—begin to die.

As I mentioned earlier, activity in any given region of the brain directly increases blood flow to that region. Now, as the brain moves on a path toward AD, these now-inactive "highest" brain zones receive much less blood. That poorer blood perfusion impairs the ability of the immunological processes of the brain to clear out the natural detritus and debris, including the toxic beta-amyloid, which is flooding out from this now-degenerating tissue. There is a very compelling literature that has directly related this accumulated debris and the failure of the immunological processes in the older brain to gobble it up to the formation of those poisonous amyloid exudations and crystals, and to the generation of those tangle-filled nerve cells.

This perspective is supported by a large body of studies that have shown that AD begins in these now-under-activated and now-blood-deprived brain areas, then spreads its poisons backwards across our great forebrain systems in the brain, where brain level after level succumbs to the consequences of reduced activity, reduced blood perfusion, compromised immunology, and a debris-filled brain.

How have drug companies tried to solve this kind of problem? They've tried to use chemicals to stimulate the clearing of debris, have attempted to apply chemicals that can result in the dissolution and removal of the poisonous crystals, tried to chemically block the process of tangles forming, and sought drugs that could regrow nerve cell connections. But none of these approaches has stopped Alzheimer's disease in any significant way.

A consortium of large companies met nearly two years ago to discuss the outcomes achieved with a number of different new drugs, all designed to stimulate the immunological system to clear up the debris in patients with AD. In every case, their treatment helped reduce the numbers of those poison crystals. However, the drug developers appeared to be genuinely surprised to discover that no AD patient was "cured." In fact, absolutely no measurable improvement in the cognitive status of treated patients could be recorded.

Why did this several-billion-dollar investment amount to naught? Senile dementia and other expressions of AD arise from a severely, progressively, functionally degraded brain. Those hand-grenades had done great damage in ways that contributed to the final period of functional deterioration. The brain had struggled mightily to retain control, but ultimately failed catastrophically. Cleaning up the debris in those thousands of little blast sites does not—cannot—reorganize and recover your brainpower by itself, because the wiring in the brains of these unfortunate individuals has been plastically revised as the pathology has progressed in their brains. No drug can fix this kind of problem. You have to learn your way out of this very deep hole. And, alas, in the AD patient, generating all of the brain wiring changes required to get out to see the light again can be a very daunting prospect.

At the same time, brain plasticity science tells us that this progression toward this disastrous outcome in most individuals who would otherwise advance to AD is probably preventable from the outset. What if these people had kept their brain machinery in better shape, so that those areas in the brain serving their "highest" brain functions were never in serious danger of going "off-line?" What if they had exercised their brains in ways that assured that those areas would always be well perfused with oxygenated blood? What if they had engaged their brains in other ways that assured that their immune systems were always operating to effectively clear out any accumulated debris and beta-amyloid in these susceptible brain regions? What if they had had simple strategies in hand by which they could assay the functionality of these brain areas, to help them maintain, through targeted brain exercises, a safe distance away from the cliff?

Strategies for delaying or potentially preventing Alzheimer's disease by working throughout your life to sustain your brain power at a high level, or by rejuvenating your brain at an older age, is an important subject for Part 5 of this book.

For further explanations and extensive references and citations related to the information in this chapter, please visit www.soft-wired.com/ref/ch26

27

THAT TROUBLESOME BODY
HOOKED UP TO MY BRAIN

A Healthy Brain Is Crucial for Sustaining a Healthy Body

Physical fitness is a major contributor to brain fitness. Its aerobic values have a direct, positive physiological impact on the brain. A person can greatly advantage their brain by maintaining physical independence as long as possible, because your ability to operate in the world contributes so strongly to that crucial real-life exercise of your brain. Not surprisingly, studies have repeatedly shown that limitations in mobility shorten the months or years that you can expect to have ahead of you before senility onset or death. Of course some of us are just not lucky enough to be able to sustain our mobility because of injury or disease affecting our body or our brain. In Part 5, I'll discuss how an individual in that situation can compensate in useful ways for that lost resource, through specific less-conventional forms of brain exercise. Meanwhile, the more fortunate majority of older people who can move well should probably be giving some thought into how to effectively exercise the brain's control of their body.

When most people think about physical exercise, they don't give much thought to the brain's role in it. In fact, your brain has been constructed to control your actions in movement, just as in thought. Your brain is begging you to engage it with a broad array of physically responding and movement-control exercise, and in the mental rehearsal that supports its refined control.

Modern culture doesn't do a very good job of helping people maintain the flexible and dynamic control of their physical actions. Humans were neurologically constructed to operate in a physical world in which every step is uncertain, every glance can be perturbed by an unexpected footfall, fine adjustments in posture are almost continuously required for movement control, almost every movement requires coordinated actions that involves whole bodies, and new physical challenges arise almost every time we take a new path across our territories. Massive natural brain exercise results from these basic physical operations.

Today, however, our cultural evolution has done a wonderful job of impoverishing the maintenance of our movement control and the neurology that supports it, by eliminating most of the challenges of our natural physical environments. Modern humans live in a paved and tiled world. We move around in a world of artificially smooth, hard surfaces, usually while wearing hard-soled shoes. We hardly ever take a step that requires any thought or adjustment in balance, that perturbs our vision in any unpredictable way, or that requires any very sophisticated control of whole-body-coordinated movement. Many of us religiously avoid prolonged physical effort. We now ride across the longer distances as largely passive actors controlling the powerful machines that convey us across them. Wagging our feet, shifting our hands along a hard surface and bending our arms at the elbows is not exactly a form of effective brain- or body-challenging exercise. In fact, our paths on foot or as riders are designed to be unchallenging and stereotypic. Our passages are smooth, our turns are gradual, there are no obstacles in our way, and our world is constructed to be almost completely devoid of surprises. We are not really required to drink in much about the details of our three-dimensional landscapes; roadmaps or verbal instructions from our machines substitute for any real reconstruction of our territories. We can be—and often are—almost brain dead, for much of the time, as we move across these flat and featureless artificial paths.

I grew up as a boy near the edge of a small town in Oregon. As a child, I spent a lot of time exploring the natural environment. The Santiam River was about a kilometer to the east of my home, and inviting woods and mountains within easy walking distance ringed my hometown. I climbed many of these hills and buttes, as well as a lot of rocks and trees that covered them. Walking across this countryside was physically and mentally challenging. It was full of wonders, full of surprises, never the same for any two passages, even when I tried to duplicate an earlier route. It was full of noises and smells and visions and feelings that spurred my brain to investigate their natures and sources. One could easily imagine that a deep understanding of one's territory would, in an earlier human era, be absolutely essential for survival.

You might want to ask yourself, "How well have I constructed a model that includes the details of the world that I live in?" Think about this for a few moments. Begin with a mental picture of the details of all of the houses, yards, road signs, sidewalks, and so forth on your street. After going house by house on your block, move that mental reconstruction three or four blocks away. Now, create a mental reconstruction of all of the details of that block. Do as complete a job in that mental construction as you are able.

Later today, walk down your street and see what details are actually there, and compare them with what you saw in your mind's eye. Modern humans will usually find lots of things that they just hadn't noticed enough to register—including things that are literally in their own front yards! Now walk that three or four blocks to that second location, and just look at what's there, in detail. If you have no surprises, and if you really had recorded it faithfully in your memory, you are one of a handful of truly exceptional modern individuals.

There is one especially important negative consequence of spending most of the years of our life in our highly predictable, structured modern environments. We've done our very best to remove anything that might actually surprise us. Surprises are very important for our brains. We're designed to give them special treatment, because in the real world, correctly interpreting them can be crucial for our very survival. The brain is sharply alerted by the unexpected by the impulsive release of noradrenaline. When the surprise is positive, our brain floods with pleasure-inducing dopamine. When the surprise portends danger, our brain shifts to that fast-action mode that can increase the probability that we can avoid or successfully confront it. There is just a little bit of surprise in every natural step, coming to us from our eyes and bodies. Thousands of times a day, the machinery that alerts and rewards and warns us is engaged. The integrity of this machinery is sustained by the release of "growth factors" that reflect the levels of its activities. Our modern, sterile landscapes provide precious little of this invigorating activity.

When we're not moving across our paved walkways and highways, we spend the majority of the rest of our day sitting. With our forearms and our hands moving out from our transfixed, disengaged trunks like little robotic appendages, this is not exactly the best postural platform to elaborate and sustain the coordinated movement control of our bodies! Our hips and back and shoulders and legs and neck are begging to be brought back into play.

Humans were constructed to actively respond, by actions expressed in our bodies, to what we see, hear, or feel. In our modern cultural state, we spend hours every day passively drinking in the significantly addictive visual and auditory and directly emotional stimulants that are delivered to us via our modern media. This leaves the body and the brain out of the equation; the brain's sophisticated action-control machinery is gaining very little from your spending all of those hours every day with it pretty much completely shut down.

The standard physical fitness training center and conventional physical therapy practices are designed to help us compensate for that lack of physical activity that is an almost unavoidable consequence of a modern life. There is a strong literature showing that good, physical aerobic exercise is very good for your brain. Unfortunately, in many of its forms, the modern gym does not do a good job of helping us exercise and re-grow our brain's control of our actions in the world. Most gyms and rehabilitation centers are still substantially brain-less. They focus on flexibility, strength, and relatively slow adjustments in balance. To the extent to which they embed movement stereotypy, or anything short of whole-body coordination, they can contribute negatively to your neurological control or recovery of movement. A large part of the problem with mobility, balance, stiffness, or agility actually lies with the controller of those faculties—the brain. While work on muscle strength and flexibility and posture are also important, they have relatively little to do with rejuvenating the brain's refined control of your movement under almost any practical circumstance. Progressive non-stereotypic exercise strategies that gradually recover precision and control, and that progress in training at speed, are important further dimensions of fitness training that the older brain needs.

Of course, there is much more to maintaining the brain's corporeal scope than its control of movement. Information from the body senses, balance organ, and vision control the adjustments in the moment-by-moment flow of blood that must be accommodated as you change your posture. We very heavily exercise this vascular control machinery in a natural environment. Although we can exercise it in a modern environment by regular walking or running or biking, our experiences are still impoverished relative to what we would experience if we actually lived our lives in nature. Many individuals do walk, run, or bike, but with earbuds in place. This means they're missing out on a crucial piece of the puzzle—because the brain benefits of physical exercise do not come exclusively from the pounding of legs on pavement and an increased heart rate. The senses must also be engaged and incorporated into the action.

There are many other influences that come from, or are expressed in, the body. A deeper understanding of the brain's role in controlling them can lead to more targeted exercise and control of these influences. Eating, drinking, urination, digestion, defecation, breathing, emotional reaction, and pain response are all on this list. All of these abilities are controlled, with feedback from our bodies, in our brains. Let's consider five different examples that illustrate how changes in the way that you live your life can influence the way that your brain can distort and reprogram the functionality of these key systems.

Dave's sore back is driving him crazy. He twisted it more than a year ago, and even though the neurologist and the neurosurgeon have both told him there is no longer any obvious physical injury, he spends a good part of every day trying to just get comfortable. "Why does it hurt so much," Dave thinks, "if there is nothing wrong with it?"

Because the brain is plastic, when pain grows, it can encode stronger and stronger pain sensations over time, despite the physical injury appearing to be healed. Scientists can record pain's slow growth in a brain, and its frequent persistence even after the original cause of the pain has evaporated. The activity in the brain regions associated with pain are powerfully influenced by the emotional context in which the pain has arisen and is sustained. In one context, the pain is strong, and can be expected to grow. In a second context, the same source of pain is weak, and can be expected to wane.

One way to illustrate this effect of context in an experimental setting is to give you a cue (for example, a red light) each time a painful heat stimulus is applied to your skin, and a second cue (a green light) at moments in time during which no painful stimulus is delivered. After you learn that red = pain, and green = no pain, I could begin to lie to you just by manipulating when I turn on the red and green lights. If you were the subject of such an experiment, you would discover that you would often fall for the lie: When you expect pain, it hurts; when no pain is expected, it doesn't hurt. People with chronic pain often learn, all too well, to expect it to hurt.

One of my favorite examples of the role of context on pain comes from a study conducted in World War II, when military doctors treating wounded soldiers gave them morphine ad libitum. During this invasion near Salerno, Italy, a timid American general led a large body of troops onshore at a basin surrounded by mountains. Success was dependent on surprise. Alas, General Lucas delayed moving his army inland while he secured his logistical support. This was a great mistake, because the German General Keitel moved mobile cannons onto the mountain ridges above the basin and mercilessly shelled the American troops down below. About half the soldiers in this hell-hole were killed or wounded. Even though many soldiers were severely wounded and could have all the morphine they desired to relieve their pain, very little anti-pain medication was requested.

Back in the war industries in the United States, another group of men of the same age, equivalently wounded in industrial accidents, were also given all of the morphine they desired to relieve their pain. They suffered terribly, and consumed the anti-pain medicine in great quantities. In Salerno, even a severe injury meant survival, and a ticket back to home and loved ones. In an American war factory, an injury of the same magnitude was a personal and family disaster—meaning no job, no money, and uncertainty about future prospects for a normal life. That difference in context, from the brain's perspective, was titrated by that great difference in the quantities of morphine that was deemed necessary to overcome the pain.

Many people are not interested in hearing that their aches and pains are aggravated by how much they dwell on them, by how often they are sure that they should hurt, or by how much their suffering might negatively impact the quality of their lives. Yet, at the same time, those nuances are complexly contributing to how much it actually hurts. While it would be foolish to dismiss the all-too-real physical suffering that can plague and degrade quality of life, at the same time it is helpful to reflect on ways that people in pain can minimize the magnification of suffering. I'll revisit this topic in chapter 34 and discuss how we might engage the brain to our advantage to try to bring our aches and pain just a little more under control.

Sandra finds herself getting just a little bit dizzy when she gets up to go the bathroom in the middle of the night. The balance organ in your body senses rapid changes in posture (for example, from lying down in bed to sitting up) and sends that information to your vascular system to adjust the blood flow to different body regions. When you sit up suddenly from a prone position, rapid changes in blood pressure in your vertebral arteries are especially crucial because you now have to pump blood to your brain "uphill" to offset the forces of gravity. The status of brain-body adjustments of blood flow as they relate to posture changes provides a good index to mark how well or poorly a person is sustaining brain health.

About a decade ago, researchers found that people who live in an area with cobblestone streets have fewer problems with these cardiovascular adjustments as compared to people who live with paved streets. They are less prone to getting dizzy when they sit up from a prone position. This is because walking on cobblestone streets provides a hidden source of effective brain exercise every day! Having an irregular surface to walk on appears to provide these people with enough exercise for the balance organ to make a significant positive difference in the health of their circulatory system. In this simple way, they are exercising a critical balance resource (i.e., the balance organ detecting rapid changes in posture) that just happens to also be crucial for controlling posture-related adjustments in blood flow. That contributes to a heavy, regular schedule of exercise for crucial brain-body control machinery.

Malik has stiff, aching joints, and feels like he hurts all over. It makes him cranky and unwilling to get out in the world, and he's lost some of the verve and vitality that he once had. Studies conducted by my research group that were designed to determine the origins of inflammation and pain in the wrist and arm associated with a repetitive strain injury (e.g., carpal tunnel syndrome) showed that the quality and strength of sensory feedback from the body influences the balance of blood flow into a limb. When the sensory feedback from this region is degraded (as it is in aging, for all of the reasons described earlier), blood perfusion in the superficial tissues to that part of the body is reduced. Because these tissues aren't receiving enough blood, they become especially susceptible to repeated-motion-induced injury, which is the basis of the symptoms of the repetitive strain injury. These studies have also shown that these conditions may be reversed by intensive, brain plasticity-based training that restores the quality of sensory information from the limb.

As people get older, they can get to the point at which they feel like they hurt all over. I believe that this arises in part because our brains receive degraded information from the body that results in negative adjustments in blood flow that alter tissue perfusion patterns, and ultimately lead to more achiness and frank pain. To the extent to which this is true, brain fitness exercises targeting the refinement and the strengths of normal somatic sensations from the skin and from deep tissues may be a path for weakening the body's aches and pains.

Bobbie is stressed out, and craves comfort food. She feels happier when she's eating, but worrying about all the weight she's gained is adding to her already high levels of stress. Brains are very good at detecting stress, and through a complicated set of processes, our brains may ultimately drive us to eat big plates of pasta or sugary desserts. When we're under stress, our brain leads us to food that's high in carbohydrates. And when we eat them, the fat that they produce actually generates brain signals that feel comforting, by reducing the chemical responses to stressors in your brain.

Isn't this ironic? We want to reduce stress because being stressed out is unhealthy. But the brain and body deal with stress by guiding you to fattening foods, so we gain weight, and now we become even more stressed.

Why, under stress, does this happen? From a brain plasticity perspective, under conditions of stress, your brain is seeking stronger chemical rewards, which translates to eating exactly the kinds of foods that generate the strongest dopamine release in your brain—the carbohydrates.

More importantly, from that same brain plasticity perspective, how can a person reduce that addictive demand? The first step in extricating oneself from this cycle is to reduce the stress load that causes the dopamine-seeking in the first place. Of course, that is easier said than done. Another approach—with the goal of tweaking the brain's chemical reward system through brain exercise—may be a more achievable and ultimately more enjoyable goal. Increasing the power of fun and joy—which are the brain's counter-balancing forces for the consequences of stress—by heavily exercising the modulatory machinery of your brain, and by living life in ways that promote that happiness, can have more lasting effects than a bag of potato chips. To put it another way, if you have true brain comfort, you won't need to eat cake to make yourself feel better.

Heather is plagued by urinary incontinence, and it's so embarrassing and inconvenient that she hardly wants to leave the house. About 40% of women and about 15% of men over the age of 60

suffer from "urge incontinence." As small children, we learn to control our urination in a process that educates the brain and spinal circuits to control the sphincters in the urinary tract and the muscles in the bladder wall to wait until the time is right. We teach the brain when and where to pee. Strange as it sounds, each time a person urinates, the brain rewards you with a very small dose of pleasure-inducing dopamine.

As we age, the brain's eager anticipation of that reward drives an incontinent individual to an uncontrollable urge to pee. Just like your brain seeks comforting carbohydrates in times of stress, incontinence may be a case of looking for a cheap thrill! There are a number of such vicissitudes of older age that we attribute to "body failure," without clearly understanding that the controller of that body—our brain—is actually a willing accomplice in that crime!

We often think of what we do to sustain our physical health as being achievable by exercising everything from the neck down. We often think of sustaining our mental health as being achievable by exercising from the neck up. Neither perspective is correct. Sustaining and controlling our brain's operations is just as important as retaining our ability to operate in the real world. And the health of that complex body machinery that controls those operations is highly brain-dependent. We need to keep it all in good condition, if we are to enjoy a maximally successful life.

For further explanations and extensive references and citations related to the information in this chapter, please visit www.soft-wired.com/ref/ch27

PART FIVE:
STRENGTHENING, CORRECTION, AND
REJUVENATION THROUGH BRAIN TRAINING

28

TEACHING OLD DOGS NEW TRICKS

How I Began to Realize that There Are Very Good Uses for a Plastic Brain

By the late 1980s, my colleagues and I had shown that the brain could be dramatically altered by specific forms of brain plasticity-based exercise. In those early studies, conducted at the University of California at San Francisco (UCSF), we showed that brain remodeling followed what scientists call the "Hebbian rule." Donald Hebb was a Canadian psychologist who had created intuitive models designed to explain how a brain might remodel its circuitry on the basis of experience. His basic theoretical principle is often stated as "What fires together, wires together." Put in other terms, according to his theoretical rule, all of those brain activities that occur together (in neuroscience parlance, "fire" together) strengthen their connections and inter-connections ("wire" together). Accumulated strengthening of these connections— and the increased "teamwork" that results from it—accounts for the improvement or acquisition, and the facile and reliable control of any practiced behavior.

Showing that brain plasticity followed Hebb's theoretical rule was important because it meant that brain plasticity processes must be fundamentally reversible. That led to experiments in which we demonstrated, as predicted, that it is just as easy to degrade as it is to improve brain function, by engaging an animal in different forms of behavior. I could improve your abilities to accurately receive and respond to my spoken words—or, just as easily, degrade those abilities—and given enough time, utterly destroy them, through relatively simple forms of brain training.

This insight led to a consideration of how the behavioral history of a brain would be expected to confer strengths or weaknesses within it. It led to our understanding that some of the major brain illnesses that commonly arise in us humans (schizophrenia, bipolar disorder, depression, obsessive-compulsive and phobic behaviors, epilepsy, acquired moment disorders, and so forth) are actually expected "failure modes" of our self-organizing brains. By that view, these conditions are not really "diseases" at all. They are simply dysfunctional situations that arise in a self-organizing brain that is driven, through perfectly natural processes, in a negative progression towards a distorted functional state.

Think about this simple example, which my colleagues and I first began exploring more than 20 years ago. Imagine a person has a job that requires that he use his hand repetitively, in a way that results in the heavy, simultaneous excitation of tactile and muscle sense inputs from that hand. Perhaps he is a graphic designer who spends all day holding a computer mouse in his hand. Because those activities from large areas of his skin and muscles are excited ("firing") simultaneously, by the Hebbian rule, every time he clicks his hand down on his computer mouse, if he is paying any attention to those feelings (a necessary condition for enabling brain plasticity), his brain will slowly strengthen the connections with one another ("wire" together) from all across his simultaneously stimulated palm and finger surfaces. The result? The detailed treatment of information from the skin and deep tissues in that part of the hand will be progressively degraded, and before long, will be useless for guiding the control of refined hand movements. In extremis, such a behavior can be expected to result in turning a hand into a clumsy, uncontrollable claw.

To confirm that this kind of learning history actually accounted for the emergence of a disorder known as a "focal hand dystonia"—a common problem in the modern work world—we trained a monkey on a similar behavior, and sure enough, the brain zone representing its hand was transformed into a very degraded processor of sensory information, and shrank dramatically in size. And just as we predicted, the monkey quickly lost control of its hand.

Focal dystonias occur very frequently in modern humans. This condition is not a "disease" or "illness," as had long been claimed. To the contrary, it is an expected mode of failure of our plastic, self-organizing brains. If you'd agree to volunteer for the experiment, my colleagues and I could easily turn your hand into a useless claw by engaging you in a few short weeks of training!

In my city, there is a clothing manufacturer with a workroom full of designers who used to spend all day moving a mouse across their mouse pads as they sketched out their clever new clothing concepts. Just like the monkeys, and for exactly the same reasons, about half of these artists developed a focal dystonia in their hands. Realizing that their designers were at risk for this predictable work-related disability, the executives of this enlightened company actually developed an educational program for prevention and employment cross-training so that if and when a hand became dysfunctional, there were other good work opportunities in the corporation for these valued workers.

Nancy Byl, a scientist who collaborated on our studies of the neurological origins of focal hand dystonias, carried our results directly into clinical practice in her physical therapy clinic at UCSF. There, she met two of the world's greatest guitarists, David Leisner and Dominic Gaudious. After achieving high professional success, both musicians developed a focal dystonia on the job. Nancy and I knew that distorted sensory feedback from the skin, muscles, and joints of the hands could only be renormalized through extensive, progressive practice. Independent of our research findings, but applying the training strategies consistent with our research, these artists drove the necessary corrective plastic changes in their brains. Both modified their lifestyle, their approach to performance, their biomechanics, and their awareness of the sensory aspects of their interface with their instrument. Both recovered, and considered their performance superior to their skills prior to the development of their signs and symptoms. They quit looking to others for a cure. Rather, they figured out that they had to accept the responsibility to change their brain, not their hand. Guitar aficionados are blessed by the complete recoveries of these two artists, which, in David Leisner's case, occurred after a 12-year performance hiatus.

With all of that in mind, it is useful to consider a more complicated example of how your plastic brain can lead you quite naturally into very deep trouble. Imagine a person with an inherited weakness that affects her ability to develop strong and reliable associations between the information about the real world and the ability to record things in her brain. This weakness in working memory and memory association degrades her ability to predict the next expected event. Compared to a normal individual, a very large number of events will be unexpected—surprises—for her brain. As I described earlier, surprises to the brain get special treatment. We release noradrenaline when an unexpected experience occurs, and it alerts our brain in ways that assure that we pay attention to it, to determine whether or not that surprise represents a danger or a potential reward for us. If the surprise is interpreted to be neutral or positive, the brain also shoots out a pulse of dopamine.

Dopamine release contributes to that little splash of curiosity or pleasure that is a frequent companion of the unexpected.

While measured amounts of these brain chemicals have necessary and positive effects in most people's brains, an imbalance in modulatory neurotransmitters like dopamine and noradrenaline can have severe consequences. If I were to artificially inject enough noradrenaline into your veins, you would become intense, paranoid, and anxious. If I were to infuse enough dopamine into your veins, you would begin to hallucinate, become delusional, and express your thoughts in a disordered way. If that dopamine poisoning was elevated enough to reach a high constant level in your blood stream, it would have another insidious effect: it would amplify the difficulty that your brain has in controlling its memory associations and predictions. With its associative and predictive powers now fatally wounded, your brain would become a "surprise machine"; it would now be almost continuously dosing you with abnormally high levels of noradrenaline and dopamine.

We call this particular failure mode of a self-organizing plastic brain "schizophrenia."

I should add that there are many other deficits that parallel this initial weakness, or that are consequences of this ultimate, catastrophic breakdown in memory control. Given this fundamental weakness, as the brain struggles to make sense of things, it begins to make many secondary adjustments. In a person with schizophrenia, many brain processes eventually come to operate in a weak or distorted way.

Because plasticity processes are, by their nature, reversible, I began to ask myself if it was reasonable to think that it should be possible to overcome a phobia or depression or schizophrenia or focal dystonia by driving a brain, through training, in corrective directions, back toward normalcy. This would be a huge positive step, because we know that treating these disorders with drugs is simply not good enough. A dysfunctional brain is not a chemical stew that is missing a key spice. Brain chemistry is extremely complicated. In a person with schizophrenia, hundreds of chemical processes have been altered by the plasticity-driven changes in the patient's brain. The notion that a single drug can provide the sole basis of treatment of all of that fouled-up machinery in a complex condition like schizophrenia is patently absurd. There is a far greater prospect for achieving neurological recovery through brain training designed to reverse the many distorting changes that are a part of the illness—and through that training, renormalizing the brain physically and chemically.

To fulfill the great potential for directing intrinsic brain plasticity to help an individual broadly "re-normalize" their distorted brain, we had to evaluate the extents to which we could drive the brain correctively, as might be required to address a complex problem like that presented by the highly distorted brain of patients with conditions like schizophrenia. With the help of world-class university laboratory scientists, my colleagues and I began a long series of laboratory studies designed to show us which neurological dysfunctions or malfunctions could actually be "fixed." Now, happily, many other scientists around the world have conducted many other practical studies on brain plasticity. At this point, we have collectively demonstrated that just about every aspect of our brain-power, intelligence, or control—in normal and in psychiatrically or neurologically impaired individuals—can be improved by intense, efficient, appropriately targeted behavioral training.

My top ten list of changes that have been demonstrated by brain plasticity research to date is as follows:

1. We can (re-)strengthen the interconnections between neurons. This synaptic strengthening is the lifeblood of brain plasticity. The re-strengthening of our brain connectivity in shrinking and progressively more disconnected older or distorted brains is now an oft-demonstrated neurological consequence of appropriate positive learning and enrichment.

2. We can improve the health and vigor of key nerve cell populations—including groups of nerve cells whose neurotransmitters enable learning and vivify memory. We know that the intensive exercise of this crucial machinery revitalizes it, grows the shrunken nerve cells, increases their energy production and utilization, enables the production of larger quantities of their neurotransmitters, and revitalizes other parts of the cellular machinery that improves their effectiveness and their chances for longer-term survival. That's pretty important, because these nerve cells control our capacities to learn or remember, and because, as we've earlier noted, the slow death of these brain centers is a major forerunner of both Alzheimer's and Parkinson's diseases.

3. We can slow the shrinking of brain centers, and in some cases increase the physical size of areas that support learning and memory processes, and we can (re)thicken targeted sectors of shrunken gray matter. That can be achieved by intensive, serious schedules of learning that increase connections between synapses and prevent nerve cells from dying. In at least some areas, like the hippocampus, training can also promote the growth of new nerve cells.

4. We can systematically increase (if necessary, recover) the accuracy with which the brain represents the information it receives from listening, vision, touch, balance, olfaction, taste, and proprioception, and we can increase the accuracy and fluency of movement and thought control, by specific forms of progressive training.

5. We can significantly improve the capacity for our brain to enduringly remember what we see, hear, feel or learn, and we can improve the strength of the associative memory processes that guide our neurological progressions in thinking, reasoning, and acting.

6. We can speed up the operations of the brain in all systems in all kinds of ways that improve the sharpness and completeness of how our brains represent and record information that we receive in fast time, or that we deploy to control our actions in fast time.

7. We can reduce the time it takes the brain to ship information around in our great functional brain systems, we can re-insulate those wires that have become weaker and less reliable in their transmission of information because it has lost insulation, and we can increase the capacity of our brains to coordinate activities across the great subsystems that separately represent different related aspects of an ongoing behavior.

8. We can improve our abilities to broaden and agilely control our attention windows and our shifts of attention, to take in more information with sharper separations or more useful integration, in vision, listening, feeling, and in our other senses. And we can improve our ability to suppress noises and chatter and distractions, to help us stay on track when our brain is searching for the right answers under less-than-ideal conditions.

9. We can re-strengthen specific skills that support our independent lives. On the side of the brain's control of our actions, we can re-strengthen and increase the security of our mobility, we can again assure faster and more reliable and more extensive responses and reactions that protect us from falling, we can improve our ability to rise up out of bed, out of the easy chair, or up off the floor. We can improve our visual skills and our speed of reactions that support our safely driving a car. We can improve our abilities to move safely and confidently in a crowded environment. And we can similarly rejuvenate long-under-practiced skills that support our independent mental actions and control.

10. We can regrow the fluent and self-confident brain operations that help us operate once again with more of our old panache and savoir faire. And we can improve the brain machinery that supports a more positive, happier life.

As my colleagues and I demonstrated that we could drive positive or negative changes in our neurology in all of these (among other) respects, I decided that it was important to determine whether our brain's plasticity changed in any fundamental way toward the end of life. More than 20 years ago, my University of California colleagues and I first tried to answer this question by studying older monkeys in which we changed the balance of behaviorally important inputs from a monkey's hand to its cerebral cortex. Would the plastic changes in the monkey's brain occur as rapidly and be expressed as strongly as in a younger animal in the prime of life?

Those studies seemed to indicate that there were no great differences between the functional and physical plastic changes induced in a monkey judged to be about 25 years of age—i.e., close to end of its expected lifespan—compared with those recorded in a vigorous 2.5-year-old adult. It was exciting to us to see that that old brain had retained at least much of its youthful plasticity!

Several years later, one of my former research collaborators, Hubert Dinse, now a professor at Bochum University in Germany's Ruhr Valley, conducted an important experiment that also demonstrated that plasticity was powerfully in place at an older age. His approach was to study rats very near the expected ends of their lives. At this point in their life cycle old rats consistently fail a simple agility test: they can no longer cross a narrow bar to get to food (that they are very eager to access) without falling off. They were now lethargic, and by any measure, cognitively impaired. On the average, Dr. Dinse's old rats would have been expected to have reached life's end in another month or two. When his research team examined the brain's representations of their movements, or of the tactile inputs from the paws that could contribute to guiding refined movements, they were found to be very grossly degraded in their old age. Their brains no longer represented information that was crucial for the control of their voluntary actions in a clear or reliable way.

Given their degraded and disorganized neurology, it was little wonder that the old rat near the end of life falls off that bar, or struggles to feed itself when that requires the manipulation of its food with its now-very-clumsy forepaws. And it is little wonder that not too many days or weeks later, the typical rat loses control of its hind legs. If these old rats were allowed to endure the obvious suffering, they would be reduced to pulling themselves around with its forelegs to gain access to food and water, with their immobile hind legs dragging along behind. This pathetic condition obviously presages these animals' death.

Hubert and his colleagues asked themselves if this sad, expected end could be altered by appropriate forms of cognitive and physical exercise. They discovered that if they intervened just one month earlier by simply providing rats with a rich variety of daily challenges that required that they intensively exercise the movement, tactile, muscle, and joint representations of their corporeal worlds, these old codgers were functionally rejuvenated. After only two weeks of such exercise, they recovered their agility to the extent that they would no longer fall off the bar. They were able to sustain the effective use of their limbs in feeding for three months longer than would otherwise be the case. Their lifespan was increased by 15%! And when Dr. Dinse reconstructed those representations of information from the senses that critically supported movement control, they were restored back to the level of a much-younger animal. This was a pretty sensational outcome for that older rat, attributable to only two weeks of fun training that meaningfully challenged the rat's brain and body.

Still more recently, several teams of scientists across the world have shown that in mice, the emergence or progressions of the pathological signs of Alzheimer's disease can be delayed and reduced in magnitude. All it takes is to provide Alzheimer's pathology-prone mice with an enriched physical environment that requires daily problem-solving in a continuously changing home-cage environment. For a human, this would be the equivalent of having to find your meals in a different hiding place every day and having to complete an obstacle course to get clues to the hiding places. Mastering those daily challenges to the end of life is exactly what our species' (and other mammals') brains had originally been designed to achieve. If that's the "secret sauce" that keeps our brain in good stead, how can we achieve it in our modern environments, without returning to a caveman/cavewoman lifestyle? That's one of the main subjects of the remaining chapters of this book.

To be fair to the science, I should also tell you that other studies conducted in cage-reared animals like rats and mice—and in some compelling cases, in humans—have documented significant differences in learning rates attributable to aging. However, I believe that those differences are substantially attributable to two factors:

- The functional decline, attributable to "disuse," of memory abilities that crucially support learning, and of the "modulatory control" machinery of the brain that critically enables learning; and
- the degradation of the brain's ability to suppress internal and external sources of distraction, which often leads to inaccurate responding in almost any real-world learning context.

We now know that the brain machinery responsible for remembering, learning, and suppressing distractors can be substantially rejuvenated by training. We now know that, at least in a rodent, those improvements in the aged animal can bestow a capacity for brain change that approaches the capacities for learning expressed in the very prime of life. Wouldn't it be wonderful if we could sort out how to achieve that for us humans? If we could, our learning and remembering efficiencies would be right back at the peak performance level that we operated with in our young adult lives.

For further explanations and extensive references and citations related to the information in this chapter, please visit
www.soft-wired.com/ref/ch28

29

"SCIENCE TO THE PEOPLE"

Bringing this New Form of Help from the Lab into the Real World

Perhaps you can sense how exciting it was for us, about two decades ago, to begin to see that brain plasticity could provide the bases for positively rebuilding dysfunctional brains—or for getting the most out of already effective ones. My colleagues and I had begun to understand that the natural plasticity processes of the brain contributed to the functional decline recorded in almost every chronic psychiatric and neurological illness, and that they doubtless contributed to the aging progression itself. We began to see that brain changes were, by their fundamental nature, reversible. We began to appreciate that the only way that a brain could recover from a long progression that left it in a deteriorated state was to "learn" its way out of it. No pill, no electrical shock, no stimulant, no new nerve cell, no stem cell, no physical exercise, no special diet, and no supplement, by itself, could rewire the brain in the ways that are necessary to achieve true recovery. We could create thought models of plausible "failure mode" scenarios that seemed to account for maladies whose origins have mystified scientists and medicine over the centuries, and that are the source of almost incalculable human anguish, suffering, and loss. We could begin to see that our science could revolutionize our understanding of our very selves. We had fresh insights into how we could get the most out of our brains to make the most out of our lives in that limited time that each one of us has on this earth. But we still faced a huge challenge: how could we bring this science into the world as effective new forms of help or personal enrichment, for the hundreds of millions of individuals whose lives could be improved by its application?

Our first instinct was to focus on the impaired and the suffering among us, because their needs were obviously the greatest. That led us to ask several overarching questions. First, what failure modes do apply for a self-organizing brain? What "disease" names have we given them? To what extents can we explain the origins of "developmental impairments" in children as a product of their plastic brains? How could we use that understanding to define an organized plan for driving the brain of such a child in a positive, corrective direction? How could we understand the neurological scenarios that lead, progressively, to an emergent psychiatric illness, or to other classes of acquired adult impairment? How could brain plasticity-driven changes account for the limitations that arise when a brain is injured or poisoned or stressed or infected or oxygen-starved? Could we prevent psychiatric illnesses or neurodegenerative diseases in at-risk individuals by increasing the resilience in their brains through intensive, appropriately targeted training? Why couldn't we just drive those reversible plasticity processes "backward" to recover more normal function in any of these populations, or dare we imagine, to renormalize—to cure—the chronic illness or the life-limiting impairments in that child or adult that we were trying to help?

During the early 1990s, all of these issues tumbled around in my mind, and in the minds of my colleagues, as parts of the uncrystallized and incomplete puzzle of delivering effective brain training strategies to the people who needed it most. To achieve any of these goals in a practical way, we knew that our training strategies had to be efficient. We knew that when we had an impaired person in front of us—regardless of their age or reason for impairment—brain plasticity-driven reversals could be expected to require a lot of difficult and complicated effort. We knew we needed to find the optimal practical strategies for driving progressively corrective neurological change, but we didn't know what specific forms this medicine should take.

My University of California colleagues and I were actively seeking a model to begin to explore these great prospects when one landed, out of the blue, onto our laps. It came in the form of an invitation to a scientific meeting organized by a Rutgers University professor, Paula Tallal. The meeting topic was "processing speed," a subject that had completely consumed Paula since her discovery, about two decades earlier, that children who operated sluggishly in processing successive acoustic inputs were delayed in language development, and struggled to read. In other words, she had observed that kids who had trouble hearing correctly would struggle in the aural language abilities in ways that would lead to their struggling to learn to read. I had rather casually told Dr. Tallal in conversation that this was a neurological problem that I thought could be fixed through intensive plasticity-based training. My as-yet-unsubstantiated scientific boasting led to her invitation.

At her symposium, at her urging, I took the opportunity to describe in more detail why and how my colleagues and I believed that the kind of problem that she had identified could be overcome by applying a brain plasticity-based training strategy. As an upshot, we decided to work collaboratively to begin constructing a set of training tools designed to correct this problem in elementary school children. While my UCSF colleagues and I committed to constructing brain plasticity-based training programs designed to drive positive plasticity in ways that would enable more accurate and higher-speed language processing in a small group of language- and reading-impaired kids, Tallal organized outcomes trials to evaluate the effectiveness of our training.

Our small team included Bill Jenkins, a research scientist in my lab, and two young engineers, Xiaoqin Wang and Srikantan Nagarajan. Meanwhile, Paula worked with her colleague Steve Miller. Together, the six of us designed and tested the first version of what would eventually become Fast ForWord—a language training program for children with dyslexia and other language and reading impairments. The first models were completed in less than six months.

Because Jenkins and I were anxious to understand what was happening in the Rutgers outcomes trial, the computer software was designed to record performance data in detail. With the help of another key technical collaborator, Bret Peterson, we wrote custom software to assure the automatic transmission of that data back to us every day on the Internet. This was the first application of automatic Internet/database tracking of trainee outcomes (a fact later acknowledged by awarded patents). As the days passed across this month-long study, our excitement grew because we could see that all seven of the kids in the study seemed to be making good progress.

When Bill Jenkins and I flew out to meet with Paula's team at Rutgers, we were greeted with big smiles. The assessors had found that all seven children had made strong gains in their aural language abilities. Almost every measured aspect of both aural speech production and reception had been markedly improved. Importantly, brain speed deficits argued to be at the core of their language had been normalized by our relatively simple forms of training. For several of these children, the training had been transformative. I still remember one small, quiet, darling, almost-six-year-old boy who had a language age of about 2.5 at the beginning of the trial. One month later, this now-confident little chatterbox was operating in language as a normal six-year-old, and now wore a smile on his face that could melt a rock!

This was one of the happiest days of my scientific life, because it showed me unequivocally that brain plasticity science could be applied and extended to help hundreds of millions of children and adults in the world—and out across the further reaches of time, to change the lives of a large proportion of countless fellow travelers on this planet—for the better. But how could these practical benefits be delivered out to all those individuals in need of help?

At dinner that evening, I argued that we scientists had to accept the responsibility for trying to figure out how this training could somehow be made available to help all of the kids—and, in extension, adults—who would benefit from its use. I argued that we'd have to write patents. We'd have to establish a company or find an existing company to partner with. Jenkins and Miller immediately signed on. Paula correctly insisted that nothing practical could be accomplished until we completed a larger and more scientifically controlled trial. We all agreed that such a trial could be conducted with still-better and more complete software controlling our computer-delivered training, over the following summer. In the meantime, assuming that our findings would probably be duplicated in that study, we unanimously agreed that I would try to define and set in motion an effective strategy for delivering this help out into the world.

When we returned to San Francisco, I found myself on a path to creating a business. It turned out to be a steep learning curve for me. I met with the UCSF chancellor, Joe Martin, and shared our results and plans with him. He organized a star-studded advisory team to meet with me—chaired by Charles Schwab himself. Schwab asked me to prepare a business plan, but as a scientist, I hadn't the first clue about how to do that! I wrote a brief scientific treatise instead. Schwab began the meeting by emphatically throwing my document down in front of me while he said, rather sharply, "This is no business plan!" He then spent the rest of the session dressing me down in front of the other advisors. What was I selling? Who was I going to sell it to? Who would control the purchase? How would I convince people to buy it? How much would it cost? I spent the better part of the meeting feebly defending my ill-formed position, while these illustrious gentlemen pelted me with Business 101 questions. Mr. Schwab and his well-heeled pals did not appear to be very impressed by my plans.

Despite my total lack of business savvy, the advisory group saw enough in the idea to tell the chancellor that we scientists just might have something that could have a positive impact in these child populations. As Joe Martin related to me later, they had told him that he'd "probably have to let Merzenich be involved in creating a company, because it probably can't succeed without him [our team]" and that "[Merzenich] will probably have to play a leadership role in the business in its start-up phase." When Joe Martin got that feedback, he agreed to make an exception to the usual UCSF policy by allowing me to take a leave of absence to establish the business.[9]

Of course the whole business hinged on the outcome of the controlled trial that Paula and her team were conducting. When that study confirmed our initial findings, we quickly wrote two papers, which were published in Science magazine. When the articles came out, the switchboards at UCSF and Rutgers were deluged with over 40,000 phone calls from people wanting to know how they could get their hands on this revolutionary new program. That gave us the additional push we needed to convince us once and for all that we needed to find a way to deliver this software out to the public at large.

Given my rocky transition from scientist to businessman, I sought advice from business and management experts. With support from friends and family, then from an investment company, we rapidly created the initial commercially deployable forms of software, and enlisted the (initially unpaid) assistance of more than 30 high-quality speech therapists to help us evaluate it. They relatively quickly returned data to us gathered from more than a thousand impaired kids. Again, universally strong gains in longitudinal assessment indices were recorded in these children from every clinic. We now knew that we had a clinically validated strategy that could be effectively delivered into the real world that could ultimately transform the lives of millions of developmentally limited kids.

As a scientist-turned-CEO, I immediately began investing in science that could extend the application of brain plasticity-based training to impaired adult populations. Research initiatives targeting the treatment of motor disorders, schizophrenia, depression, traumatic brain injury, and other important human problems were begun. Patents were written. Prospects for great future business appeared to be wonderful. More importantly, prospects for transforming the landscape of brain therapeutics looked bright, and it seemed that the possibilities for helping people in need were almost limitless.

---

[9] That business evolved into the Oakland, California-based public company, Scientific Learning Corporation.

In the meantime, I made preparations for leaving the business in capable hands as I returned to my laboratory at the University of California. The first step was to recruit a CEO to transition the software from a handful of clinics out to the world at large. That new executive almost immediately determined that the company—now called Scientific Learning Corporation—had to "focus to win." Non-child research initiatives were abandoned, and the company directed all of its sales and distribution efforts on the large-dollar sales that could come from marketing to American school districts.

I was terribly frustrated by this direction. First of all, the plan meant that the software would only be sold through high-cost contracts to the limited group of school districts who could afford them, or with relatively expensive software for individual children that could only be provided by a family engaging a professional therapist. This meant that access to training would necessarily be expensive for everyone who received it, and therefore unavailable to the majority of American kids in need of help. A focus on American schools as opposed to the more global approach I had envisioned further limited that population of kids in need that our approach could reach out to help.

My colleagues and I also knew that our science could be applied to help many other child and adult populations whose neurological problems went far beyond struggles in language and reading in the schoolhouse. The preliminary work that had been accomplished in more severely developmentally impaired, psychotic, brain-injured, movement-disordered, and normally aging individuals, and the patents that had been awarded that seemed to open up these far-wider prospects, soon lay dormant. To me, this meant that many, many millions of individuals who could potentially be transformed by our science were going to pay a price because of our inability to get our help out to them.

After nearly a decade spent trying to influence Scientific Learning to pursue wider interests, I finally negotiated an agreement with them that would allow a new company to apply their patents—the patents my colleagues and I had initially written on our company's behalf—to help adult populations with cognitive impairments related to normal aging and other conditions. With that agreement, Posit Science Corporation was founded. It immediately directed its sights on adult brain health, and toward the development of brain plasticity-based treatments to provide remedial help to apply to a variety of psychiatric and neurological conditions. As the business focus once again began to narrow, I decided it was high time to guarantee that there would always be doors open to studying the application of brain plasticity-based training for a variety of conditions. With that goal in mind, a research arm of Posit Science was spun off as a separate company. The Brain Plasticity Institute was founded in 2009. Its goal is to conduct research and to create practical brain plasticity-based strategies that can potentially address a variety of human impairments and maladies that would otherwise receive little scientific attention.

These efforts to deliver science-based therapeutic tools out into the world have been richly rewarding. The first kids who trained on our reading impairment software are now of a college age. Recently I learned that one wonderful young child in that first study is now pursuing her PhD in psychology! Many thousands of parents and grandmothers and adults and therapists and teachers have written to Paula Tallal or to me, or have told us in person how the programs that our teams created have literally transformed their children's (or their) lives. What a great bonus it is to hear such stories for inveterate lab rats like me and Paula!

The effectiveness of these programs has been repeatedly confirmed in scientific studies conducted in many hundreds of clinics and schools and university laboratories in the world. Additional studies are ongoing at several hundred sites. Whenever a scientist has appropriately designed and conducted a longitudinal brain recording or imaging study using these tools, they've found that the training that we developed—designed for whatever purpose—has driven the patterns of responses in the brain to change, in the predicted corrective direction.

Over the next decades, I predict that we will all witness the rapid maturation of this therapeutic field, as hundreds of brain plasticity-based tools developed to reverse the plastic changes that limit the performance abilities of patients with many classes of neurological and psychiatric impairment and disease are brought out into the world.

For further explanations and extensive references and citations related to the information in this chapter, please visit
www.soft-wired.com/ref/ch29

30

REORGANIZING YOUR LIFE
WITH A BRAIN FITNESS GOAL

Daily Activities that Contribute
to the Maintenance of a Healthy Brain

At this point you may be starting to realize that the way that you have been operating in life may not be optimal from a brain health perspective. How could you reorganize your everyday life in ways that assure brain health improvements and a higher probability that you don't squander those last decades of your life as a distorted caricature of your younger self—or as a victim of a near-end-of-life catastrophe like Alzheimer's or Parkinson's? How can you improve your neurological capabilities in ways that advance the quality of your performance on the job, in those avocations that so greatly enrich your life, or in your social community? How can you hope to thrive, if fate is making you shoulder special neurological burdens on your life's path? What could you do to help assure that the person you are continues to flourish and grow? The first thing that I recommend to begin that growing or restoration process is to start by spending some time in an online brain fitness center like BrainHQ from Posit Science (or at one of our partner's brain training sites, found at AARP and Easter Seals.) My colleagues and I designed the exercises for just this reason, and have scientifically documented their benefits in dozens of peer-reviewed studies. I shall explain how these computer-delivered exercises are designed to assure that you have rapid, positive gains or recoveries in a broad range of key neurological performance abilities—whatever your initial functional status—in the next chapter.

At the same time, you should now understand that if your brain was doing well in the right ways because of how you had been living your life—if you had worked with reasonable consistency following life strategies that contributed to your ongoing brain health—or if you are at a young enough age now to follow those life strategies, visits to a brain fitness center may never be necessary. After all, we have all met the rare older-aged paragon who is exceptionally energetic, productive, and competent almost to the very end of their time on earth. While such individuals commonly do not really understand how their life strategies actually account for their good fortune, they provide clear testimony for the rest of us that there are strategies by which we can lead lives that can sustain our brains in good operational shape out to a ripe old age. The big question is: How can we organize our lives in ways that will not merely sustain—but also grow both our competencies, and that very important person that lives inside our head?

As you begin to develop a personal plan for strengthening your neurological faculties or improving your brain health, especially if you are carrying special neurological burdens with you in life, I suggest that you consider going to a professional therapist to gain some understanding of the state of your overall brain fitness. A basic but more limited strategy for assessing your brain health is provided at BrainHQ for free. While it cannot really substitute for valuable expert assessment and advice, by its use you can begin to calibrate how you're doing in comparison with other citizens of your age in general good health, or with a similarly burdened neurological history.

Special BrainHQ challenges and courses are also designed to guide you in your adoption and use of brain-changing behaviors that can potentially increase your resilience so that you do not so readily succumb to neurodegenerative illness. Many of you carry special risks for collapse into Alzheimer's, Parkinson's, or other end-of-life disasters. It is especially important for you to assay how the vicissitudes of your life may have impacted your brain health status. As you work hard to reorganize your lifestyle with clear brain health objectives in mind, you are encouraged to revisit the BrainHQ site to re-evaluate yourself. In this simple way, you can determine whether or not your own new brain plasticity-inspired self-help strategies are actually working to improve your brain health.

There are two other important aspects to reorganizing your life with a brain fitness goal. First, you should stop or substantially minimize all of those negative-learning behaviors that are likely leading you down the path to neurological ruin. If you're a typical older individual, you've already adopted a lot of bad habits that are tilting your balance between staying healthy and falling apart, to favor the latter outcome. I hope that after reading this book you are going to try to tilt things back at least a little more in favor of your brain. The bottom line: Live WITH your brain. To the extent to which your daily habits and activities can be performed unthinkingly with minimal effort and little serious attention, you are slowly taking your brain offline—in the long run, permanently. Every moment that your brain is offline, you are just a little less alive. Just as importantly, you should try to challenge your brain with NEW experiences and learning. Working with seriousness of purpose to acquire or to improve a new ability every month or two or three, from this point forward to the end of life, would serve you very well. Your learning machinery will more likely remain in good stead, and your life will be greatly enriched by all of the wider understanding and physical and mental exercise that necessarily come from all of that further personal development.

A second broad goal should be to adopt plasticity-based brain fitness principles as you operate in your everyday life.

Believe in, and follow these eight general brain fitness rules:

1. As you take on new activities and acquire new skills and abilities, develop the habit of working on the level at which you make steady, measurable progress.

2. Count every little indication of progress as success.

3. Don't forget to reward yourself, in your mind, for those growing achievements.

4. Try to work on a demanding level, at the cutting edge at which you'll improve your abilities faster—always remembering the simple rule: If it doesn't matter to you, and if you don't have to try to succeed, nothing much will change in your brain.

5. Find exercises that are inherently satisfying and rewarding. The more positive good feelings you get from your activities, the more your brain has been turning on its "Save It" machinery.

6. The best kinds of activities are those that encompass new learning and that demand that you pay attention to the details of what you see or hear or feel or smell and at the same time involve grander and more complex planning or performance challenges as you progress.

202

7. Try consciously to improve whatever skill you're trying to advance at speed.

8. Skills that have learning progressions that can never be completely mastered are always a good bet. Beware of slipping onto an easy path that levels out at a mediocre performance level. You're making the most real progress when the task difficulty always requires a challenging level of performance. Making some mistakes, and working at a task at which you can always see—and can always realistically strive for—just a little more improvement indicates that you have the difficulty level set about right.

With these guiding principles in mind, how about some specific examples of new skills that might be particularly useful for maintaining and growing brain health? Of course, each of us has very different specific interests and learning possibilities. No one else can design a brain healthy lifestyle for you. At the same time, it may be helpful to consider a few examples of how a more brain-healthy life might be organized.

One thing you can do is adopt daily activities that have the potential for further restoring your brain's peak operational abilities as they apply for your everyday life. For example, to improve your operations in language listening and language usage, sign up for an adult education class, or for a lecture course at the local college or junior college. Be a careful listener in class. Take notes. Focus with intensity on class discussions. Do not just sit and be a bump on the log! A bonus comes from meeting new, interesting people in a socially accepting setting. Work hard to make friends there. There's a lot to learn, and a course of continuous learning marked by close, careful listening in a positive social environment is always a healthy enterprise! The key to brain change is close, serious, highly attentive engagement at a level on which you are continuously challenging yourself.

Study a new language, and master it at a usable conversational level. Work on accurately receiving—and on accurately producing—word sounds for at least 10 to 30 minutes every day. Put your learning on a schedule and take it seriously. Listen intensely as a habit, in language class or through an audio class. Work hard to precisely duplicate what you hear from a native speaker or from the audio to produce the sounds of the second language both with high accuracy—and at its normal speed.

Develop a habit of careful conversational listening. One strategy might be to test how much you remember about every conversation in person or on the telephone, soon after and again a few hours after that conversation has ended. This new habit can have two advantages: it promotes careful listening to the details of what you are hearing, and it helps you remember what you have just heard. Your goal is to slowly learn to remember exactly what you just heard, with progressively greater accuracy, in progressively greater detail, in ways that support your remembering it, over progressively longer spans of time.

If you have an interest in music, rekindle it through careful listening. Try to really get into the details of the music. See if you can gradually reconstruct the orchestration in the musical arrangement, and the nuances contributed by the musician(s)—or for classical music, by the conductor. Test yourself after hearing each piece of music by seeing how much of it you remember in detail, or by how much you can accurately track (with your voice; in your mind) as the music goes flying by—or in your mind, an hour or two later. It is important in such experiences that you strive in each listening experience, to hear and remember and predict just a little more, from each listening cycle, than you had ever heard, remembered, or predicted before.

Perhaps you would enjoy learning to play, or would like to update and improve on your earlier performance abilities on a musical instrument. Harmonica, piano, guitar, ukulele, tuba—it just doesn't matter! Focus on accurate listening and performance precision as you play. Musical performance exercises reading, listening, fine and high-speed manual control, and often, other special oral skills. If you play with spirit—and why wouldn't you?—you're engaging emotion control brain hardware as well. Closely attend to, and work hard to improve just a little, in all of these dimensions of performance, every practice day.

Or, just as usefully, learn to sing or work on improving your singing voice, and develop or recover your ability to accurately follow all of the twists and turns of a progressively more challenging musical score. Your goal is to slowly develop the ability to sing just about any song from an increasingly reliable and expanding song memory, in any tempo, at the drop of a hat. Singing can be still more enjoyable when you do it with others—and there are few other places quite so wonderful for again growing your social cognition abilities than a choir.

Find a volunteer position in which you can use your language skills in interaction with other people. As an example, there are great school volunteer programs in many communities that provide opportunities for you to meet and communicate with new people of all ages. Children provide a wonderful source of serious challenge in listening, both because they can jabber on at high rates and because their grammar might be just a little less developed—and yet, perhaps a little more modern—than yours is! Again, your training task is to develop accurate listening skills with which you can understand almost anything anyone is saying, and to educate yourself about the ever-changing social nuances of the kids who are sharing the planet with us.

Incidentally, if you have finally been convinced to get that hearing aid, try to forget that hearing loss. It's no excuse. Turn down that TV. Start really listening again, with a brain that is trained to take full advantage of that new gift of better hearing. That doesn't mean that you're going to hear everything with it. It just means that you're going to hear and understand more than you have in a long time. And congratulations, if you're like me not so many months ago, for finally recognizing that you really would benefit from that hearing aid.

Talk more to your friend or mate or daughter or sister or brother about something that really matters in life or in your world, every day. Listen to them, every day. Really listen.

Different challenges must be taken on to improve your everyday operations in vision. Jigsaw puzzles represent one simple, classical form of challenge that, in principle, should be good for your brain. If that sounds a little low-tech, let me remind you that doing a jigsaw puzzle requires your close, focused attention, and that you must make fast decisions based on shape and color and visual textures to be successful. Your success depends on mental rotation of puzzle pieces and you must continually shift your visual attention from the detailed piece (engaging local attention) to the wider picture (activating global attention). Finding a correct piece is substantially rewarding for the brain. A goal is to progress, puzzle by puzzle. That can be achieved to some extent by simply completing progressively more difficult puzzles. More useful still, try to master the ability of applying different search strategies—dependent upon a color, a form, a specific color in a specific location—with a slow increase in the number of strategies and locations that your immediate search is directed toward. Also make a conscious effort to ever-so-slowly increase your speed. Jigsaw puzzles represent the kind of unmasterable exercise that you should be able to get better and better and better at, if you really try. Your visual brain would appreciate it.

Painting or other arts offer many of the same multimodal virtues. In vision, the sculptor, potter, painter, furniture maker, wood turner, jewelry maker, glass sculpture, or etcher is continuously shifting attention between fine details and a grand perspective, both in their mind and in their actions. As they master their abilities, they must evolve increasingly refined hand and body control to achieve an increasingly aesthetic visual and practical outcome. Take your continuous improvement at any such activity seriously and your brain will thank you for it. If you do it just for fun, then your brain won't care very much about your spending all that time piddling around.

Tennis or other games with the same ball-in-motion challenges (badminton, ping pong, catch, certain Wii games, basketball, and volleyball, among many others) put your visual reception machinery and action-control machinery in motion simultaneously. Games that require fast visual tracking, that drive your brain to rapidly move your eyes, and that lead to fast and highly flexible motor responses are very beneficial for both the brain and body! The exercise they provide for your neurological control of balance and posture is an important second benefit. One way to slowly progress in ability at a new game like these examples is to find a matched, competitive partner to learn with. Don't forget—in this and in all other recommended brain exercises—the special value for your brain of playing a game or two almost every day.

Just as in listening, when you go for a walk, have a social visit, go shopping, or have any other experiences away from home, you should try to develop the habit of reconstructing all of the things that you experienced on that outing, and work to reconstruct those things that defined specific, memorable scenes. Such practices will increase your level of focused, accurate vision, encourage you to put your eyes in motion so that you really examine the world in detail again, richly associate what you heard and felt with what you saw, and exercise your ability to reconstruct those scene details as a part of their elaborate whole. If you sit down and draw or summarize what you've just seen shortly after it is no longer in view, that will help you remember it over the longer term—and more importantly, provide a basis for your determination of whether or not you are slowly improving in this ability. I like to create these reconstructions in my mind, play them backwards and forwards, and self-assess whether or not I'm improving at my memory of immediate-past events over time.

I am improving my memory, even though my 72nd birthday is just around the corner. Just like you can.

When you're at any social gathering, be an attentive observer as you hear individual names, and see how many names you can recall after the party is over. Remember that recalling names is a matter of both visual and listening focus, and your ability to record those name-labeled faces. You need a sharp visual and listening record to flawlessly associate one with the other.

There is a Tibetan Buddhist practice in which the meditator intensely masters a complex image of the Buddha, on a level at which they can ultimately reconstruct every detail in their mind. With that mastery, the meditator is drawn into, and can facilely move around, this complex mental reconstruction. Mental exercises in which you record the details of and manipulate what you "see" every which way (in your mind: what would that flower look like if I were looking at it from above, from the back, in the center of a bouquet, or if every other petal were missing?) Most of these mental manipulations (and many others of this sort) can be verified by further observation in real life.

Try playing the "how many things fit into this category?" game. As you arrive at an idle moment in your day, don't grab your smart phone and start mindlessly scrolling. Don't just space out. Instead, mentally review how many things you can think of that have the same label, or that relate to that very thing that is in front of you. As I hold my toothbrush in my hand, I might review my long history with toothbrushes, what colors I have owned and favored, and given their past, what we might expect toothbrushes of the future to look like. Or I might tabulate in my mind all of the things that have been invented, like the toothbrush, that are designed to improve my dental hygiene or health. Or I might think of what defines the boundary, in my mind, between toothbrushes, or dental and non-dental hygiene. I can, of course, play this game with anything that I see or do, any old time in any day. Why not do it a time or two or ten, every day?

It is important to remember that quiet moments spent in such exercises should not be allowed to drift too often into unstructured mind wandering. Your goal is to re-engage, re-structure, and re-strengthen good habits of thought control, not to exercise meaningless, purposeless, unstructured mental idling.

Play the "how many ways?" game. The game of Boggle is an example of this kind of game, but you don't even need to go out and buy a game to practice this habit. Simply think through the alternative ways that you can answer a question or solve a puzzle that sits before you. Since I'm going to visit Sally and have been there 50 times or more, always driving by the same route, what other ways could I take to get there? The first part of my game might be to take a different path to get there every time I go there, until I exhaust those new possibilities. The second part of my game might be to draw out on paper or in my mind—every so often—how many different paths I have discovered.

Really master the use of modern electronic media, and use it to your brain's advantage. As long as you understand that your smart phone or smart pad or computer is a tool and not a brain substitute or a life substitute, the learning experiences that modern tools can provide can be powerfully enriching. Smart devices are also wonderful resources for helping you initiate new learning, and for informing you about the many new opportunities for personal growth that are out there for you in your own community. Being "modern" to this degree can be a great help for your development of the kinds of new life strategies that meet your own special life history and needs.

Learn the tango. The tango is all about many variations on a theme definitely not by Paganini! Okay, it doesn't have to be the tango. It's just a good example of a dance (or other motor activity) that starts simply and gets more and more sophisticated and complicated, and requires more and more movement variations and adjustments in balance, as you get better at leading or following the complex movements of your partner. When you are really good, you have acquired many ways to move to the same rhythm and have done a great job of training your active vision and posture and balance in the bargain! And remember that no matter how good a dancer you are, you can be a better one tomorrow.

Find a partner and play progressive games in which each time you or your opponent(s) make a play, the logic or circumstances have been changed by that play. Most card games follow that principle. If you enjoy playing cards, purchase a book of different card games and learn a new one of this nature with your partner every few months. Never stop learning new ones.

Work every day to weaken the tendency to grow your egocentrism and the forces that drive you to grow your selfishness. Be self-aware. One way to help achieve this is to set aside a time every day to fill out a personal mental checklist that has a few entries that encompass your personal variation of egocentrism (we all have one). After you score how you are doing on each virtue or fault on your own check list, immediately take each one of those scores up at least one small step toward being a better, stronger, more generous and outward-directed, and less self-absorbed individual.

With all of those examples in mind, I hope you now have a better understanding of the kinds of exercises that could help your brain—and you—be a stronger, more capable, and happier person. While you are organizing a program to fit your own personal schedule and your own life's interests, I would also suggest one more thing: stop telling yourself that you're doing OK, when you actually don't know how well you are doing for your brain. Get an expert opinion. There are many thousands of expert professionals out there in the world—including some in your neighborhood—who could help you understand how you actually measure up. And it's just not that difficult for you to go to BrainHQ for a little free self-administered self-evaluation, especially knowing that whatever you discover about yourself there is private and confidential.

There are many situations in which we ask technology to deal with those simple weaknesses that show up in life, when that means depriving your brain of very important forms of exercise. In these cases, technology may be hurting our brains rather than helping. Before you check the Internet for an answer or solution, try reasoning in your mind. Retain your mastery of the environment in which you live your life. Remember that the loss of the ability to find your way in the world, all on your own, is an important harbinger of problems to come in life. You don't really need that GPS to find your way home, and mental navigation is an excellent exercise for your brain. Use your own wits to sort through all of the possible answers to what is puzzling you. Your brain lives by its wits. Technology provides innumerable great blessings in our modern lives—but only when we put it in its proper place, in your toolbox.

My final piece of advice about reorganizing your life with a brain fitness goal is to seek and spread joy. I know that this is easy to say, and not always easy to accomplish. But understand that as your brain gets healthier, that health will be manifested to a very significant extent by a greater capacity for having fun and enjoying life. There are very good reasons to anticipate such changes. Almost every exercise strategy that I have described should add to the power of the brain machinery that controls your brightness and pleasure. As your faculties improve, so shall your self-satisfaction and confidence. Find yourself doing brain fitness-directed things that you love to do, and you may find that they will love you back!

Your job out in the real world is to work actively to grow these precious neurological resources, to make certain that they are more richly incorporated into the fabric of your daily life. The bottom line: Refocus and re-intensify your active hearing, seeing, and feeling. Re-engage with life again, with vigor, seriousness, and challenge. Nurture behaviors that are demanding on every level of perception and cognition—from addressing details of sensation and perception through complex levels of reasoning and planning. Learn to learn again. Celebrate every small step in progress, because small steps can lead to big achievements and the pleasure that accompanies them. Avoid the effortless path. Stop hiding behind the mindless, brainless, struggle-free behaviors that you mastered in your younger life. In other words, stop going to such great efforts not to engage in real life! Grow again, in your everyday activities, by improving old abilities and by developing new ones, by leading a life more richly supplied with interesting experiences, and by adopting a positive attitude and reclaiming your thirst for joy in life.

If you follow this path, there is a good likelihood that you can look forward to a better, happier and far safer life.

For further explanations and extensive references and citations related to the information in this chapter, please visit
www.soft-wired.com/ref/ch30

31

PROGRAMS FOR BRAIN REJUVENATION
AND BRAIN RECOVERY

Features of Effective, Internet-Delivered,
Neuroscience-Based Programs Designed to Grow,
Rejuvenate, and Recover—then Sustain—Brain Health

My mother-in-law Marge had the great misfortune to be riding in the car when her friend Marian, looking away in conversation, rear-ended the car in front of them. The air bag did not deploy and Marge's head slammed into the windshield. A neurological examination and scan did not reveal any immediate problem beyond the large bump and bruise on her forehead. Three months passed before symptoms of cognitive loss became prominent enough to prompt Marge to return to the doctor. Now, the doctor found diffuse damage attributable to an extended period of subdural brain bleeding. Marge's cognitive losses grew over the following several months, despite two surgical attempts to relieve the pressure and stop the slow seepage from what turned out to be two substantial brain bleeds.

Marge was the first person to be an experimental subject on the scientific path to creating the first version of what was to evolve into Posit Science's BrainHQ.[10] When she began this brain plasticity-based training program, it was obvious that this wonderful mother and grandmother was in imminent danger of cognitive collapse. That status was confirmed by standardized behavioral testing administered by a psychologist to assay her cognitive abilities. They showed that Marge was operating cognitively at the fifth percentile for all individuals of the same age. One year earlier—and throughout her younger life as a nurse

---

[10] See Posit Science's www.BrainHQ.com.

and mother—she was an exceptionally intelligent and effective individual. Now, a year later, when 95% of people of her age were out-performing her in a wide variety of memory and cognitive tasks, Marge was in very serious trouble.

Doctors often call this level of "benign" memory and other cognitive loss that threatens personal independence "mild cognitive impairment" ("MCI"). Who are they kidding? When it's happening to someone you love, there is nothing benign about it.

When Marge began working one hour each day at intensive brain-training exercises, we could all see that she was rapidly changing for the better. She worked diligently over a six-week period, for a total of about 30 hours of brain training. When she was re-tested after the training program was completed, her scores now placed her memory and thinking near the 60th percentile for individuals of the same age. Another term for this is "normal."

It should be noted that this is a strong advance for someone with losses of this severity who has undergone this limited amount of brain training. At the same time, many cases of strong improvements like Marge's have been recorded after completion of targeted, scientifically designed brain training programs. The brain training Marge did was a very early version of what we have now created, and only a fraction as sophisticated in design, delivery, and testing. But the improvements that my dear mother-in-law made were an inspiration for me and my colleagues on the BrainHQ team at Posit Science to work hard to create other adult forms of effective training programs designed to empower—and if necessary, contribute to the rescue of—adult individuals who could benefit from this kind of help. For Marge, the benefits were substantially greater. While her physical health failed her just two years later, she fully enjoyed those years as a smart, happy, sassy, loving, and independent mother, grandmother, and friend. Even as her body started to trouble her, her brain remained sharp and productive.

It should also be noted that recovering Marge's life with this efficiency and power would be difficult if not impossible to achieve through any mere lifestyle changes. There are certain situations in which there is simply no substitute for brain plasticity-based training exercises.

Before I begin describing how brain plasticity-based programs have been designed and are implemented—and explain what they can and cannot do—I want you to know that the next pages of this book may sound a little like an advertisement. But my own motive is a simple one: these programs represent the grand product of my life's work. My colleagues and I have spent many years working to create practical, effective strategies to help people. Here, I unapologetically explain the virtues of these programs to you. I describe them in a positive way, of course, because from my own scientific perspective, they represent the state of our art.

What are brain plasticity-based training programs designed to do? How do they work? Hopefully, you now have a pretty good notion about the general neurological and behavioral targets of the kinds of computer-delivered brain fitness exercises that might benefit someone like you—or someone that you care about who could benefit from this kind of special assistance. Exercises are designed to serve three general purposes. First, for some people, like Marge, the overall goal is to train the brain in ways that can drive it correctively to recover from a psychiatric or neurological affliction. Such a problem may result from any one of a number of ways in which the brain was disturbed, poisoned, infected, or injured, or it may result from a relatively serious level of disrepair that can arise in normal aging. Brain fitness recovery programs attempt to achieve a level of recovery by reversing, insofar as is possible, those brain changes that have contributed to the malfunctions and dysfunctions that account for each individual's behavioral limitations or distortions.

Second, there are people who may have escaped these insults to the brain and who wish to properly focus on sustaining their operational capabilities in ways that help them thrive. The primary goal that we all share is to maintain our basic brain health in ways that assure that we are protected from a premature decline as we age, and that increases the likelihood that we can retain our happy independence and continue to grow and thrive to the end of our physical lives. Just about every one of us, at some point in life, has to think about taking on this important brainwork, because we all aspire to having our brain last as long as our physical life.

Finally, there is a third group that is motivated to improve their neurological competencies in ways that help them excel in their work, their hobbies, and most of all, in their lives. "Getting the most out of your brain" is another way of saying "getting the most out of your life."

How could we address the different needs expressed by the different goals of all of these groups, not to mention meet the needs of each individual who comes to a brain fitness center with their own special issues and goals? In fact, there are common neurological processes that are core brain assets that almost everyone can benefit from working to improve on—and they're almost always all improvable to a significant extent, regardless of an individual's neurological starting point.

What's on this list of core brain assets?

1. Alertness and focus. An increase in your level of alertness and focus is always a direct target of effective computer-based brain training. As I've explained, your brain has a "dimmer switch" that controls how strongly your brain is engaged in any demanding task. There are not too many brains on earth that would not benefit from having just a little more spark.

2. Positive mood. When brains struggle, they commonly degrade the way that they control your positive mood. A less joyful life foretells problems with learning and remembering—and most importantly—for happiness itself. As I've explained, dopamine release is a key agent of joy and change in your brain, and most older and impaired (as well as many younger) brains release far too little of it. Almost all of our brain training exercises   target the improvement of positive mood. We try to accomplish that by adding triggers that engage the rewarding dopamine and arousing noradrenaline systems in each brain exercise. By that strategy, a few minutes of daily work provides your happiness machinery with more actual exercise than it would ordinarily get over the entire remainder of your day. Most brains greatly benefit from the strengthening or revitalization of this crucial machinery.

3. Learning and remembering. Improving your level of alertness, exercising your attention control, and strengthening your positive mood all contribute to strengthening your learning and remembering machinery. There is, of course, more to effective learning and memory than keeping those "on" and "off" and "save it" switches in good condition. Still, if those switches are working sluggishly or weakly, your ability to effectively recover or thrive or excel shall be more limited. We try to incorporate exercise of all of the complex neurological assets that contribute to the health and power of this learning-control machinery as elemental training targets for every brain fitness training program that we have developed.

4. Accuracy. Now that we have your learning and memory machinery working in a brain that is hopefully more alert, attentive, and positive, we can begin to think about strengthening other key abilities that can almost always be refined and improved. The first of those is representational accuracy. As older brains get progressively "noisier," the accuracy with which they represent the details of things slowly degrades. Needless to say, an accurate representation of those details is crucial for correctly interpreting and remembering them. Even during your life's era of peak performance, representational accuracy can almost always be significantly improved. Improving that accuracy through progressive training provides a primary foundation for growing, elaborating, and recovering all of our higher brain operations. Because it sets limits for all higher abilities, issues of accuracy are a prerequisite for cognitive training in almost any program that can effectively help you grow your brain-power. If you've ever played "brain games" on the Internet, you may have noticed that most of them focus on training you to exercise your lost abilities directly—for example, they engage you in practicing remembering or thinking or problem solving—without paying very much attention to the quality of the information that your brain has to use in order to remember, think, or problem-solve. Most people don't instinctively understand that they must also do a lot of work on improving their brain's accuracy for seeing, hearing, or feeling, or else most of that memory or problem-solving practice will be wasted. We have evolved strategies that are designed to help you rebuild or further grow these critical perceptual foundations as an important prerequisite for memory and related cognitive recovery and improvement.

5. Brain speed. I've mentioned several times that old brains slow down for a variety of different reasons. Brain speed is almost always in play in useful brain health exercises; faster brains are, in neurological terms, better at getting things right, which, after all, is the real goal of brain health-directed exercises. Speeding up your brain is pretty complicated because a number of different neurological factors contribute to it. To achieve the greatest improvements or gains, all of them must be improved. Moreover, driving your brain to go faster only makes sense if it is achieved with retained—or still better, growing—accuracy and reliability.

In practice, we almost always train by strategies that assure accuracy at speed. For the brain to operate faster, we have to engage it in ways that increase the salience of representation of the fast-changing details of what we sense, or of the neurological activities representing our memories and thoughts, or that control our actions. Speed is critically dependent on the sharpness of the brain's responses to all those little details.

Effective brain training exercises must also ultimately drive your brain to re-insulate its wiring because that is another key to recovering brain accuracy with speed. We now know that increasing the coordination—the "teamwork"—of neurons in those hundreds of millions of tiny micro-computers in your brain induces positive re-insulation of a brain's wires. That re-insulation increases both the speed and reliability of the brain. Once again, strategies designed to improve this key physical asset are built into almost all of the training exercises that we have constructed.

6. Rapid sequencing and prediction. When brains are sluggish, their representations of any given moment or event can spill over and contaminate the representations of the next events in sequence. We commonly want to make sure that the brain is doing a good job at controlling rapid sequencing and prediction, and in every program suite, we invariably design exercises that help you sharply distinguish between—and keep separate, and in order—rapidly successive events and actions. Representing your streams of successive neurological actions in crystal-clear forms also helps assure that the brain is improving its crucial powers of prediction. Your brain's stable and fluent operations are heavily dependent on these syntactic and cueing processes. Just as in a Google search, you can go a lot faster across your neurological landscape when your brain accurately guesses what it should look up or what should happen next!

7. Suppression of noise and distractions. As you struggle to follow a conversation or rapidly understand a scene in front of you, you are often bombarded by internal and external noises and distractions. A greater susceptibility to their interfering powers is a near-universal problem of any struggling brain. Reducing the interfering influences of distractors is a key for getting the most out of even a high-functioning brain. The exercises make you work on improving this key ability at some level, in many of the braining training program suites that we deploy.

8. Fluency. Having a brain that can operate with speed and accuracy helps us enhance the great gift of fluency. At the higher end of our training progressions, we commonly exercise the brain's fluent actions to assure that a trainee is translating all of their gains in accuracy and speed and sequencing into racing with aplomb across their cognitive and action-control landscapes. If we can get a brain to double its efficiency in recording information—to consider one example that is achievable through brain training and for which efficiency really matters—the individual who benefits from this training would be able to grow their knowledge about their worlds twice as rapidly as was the case before training was initiated. That would enable many individuals to reverse their backward progression, turning their brains (themselves) back once again onto a path of empowerment and growth.

9. Navigating in time and place. We live in the dimensions of time and place, and strengthening the construction, manipulation, and recording of information in these terms is a key neurological training target. In a healthy brain, their manipulation in thought is just as important as is reconstruction of events in real time and place. All of these aspects of complex scene construction and analysis, and of way-finding, are exercised in our training suites.

10. People skills. Our happiness is inextricably tied to the nature and sophistication of our social interactions. Scientific studies have repeatedly shown us that people skills deserve their own special emphasis in brain strength training. Training individuals broadly in ways that markedly improve all of the fundamental abilities that we have just described does not always equate with a strong improvement in their quality of life. The reason? Happiness can still elude you if you are not a healthy, interactive player in the social game of life. Few things can pay off more richly in training than the growth or recovery of your powers and understanding in social cognition and social control.

11. Flexible intelligence. Finally, all of the exercises at BrainHQ are designed to collectively improve your flexible intelligence. An alert, focused, motivated, empowered-to-learn, faster, more accurate, more reliably predicting, more powerfully reconstructing, more fluent, less distractible, and more socially adept brain will be that much smarter! To further assure that you apply these newly grown assets in your daily life, we commonly extend training in ways that help you directly rehearse adaptive mental strategies that are in play in your real world. It is important for you to rapidly adapt your behavior, given newly anticipated or unexpected information, or when any decision changes the logic that applies for that next decision. Our real world is full of successive, additive nuances and the unexpected; the control of our actions and thoughts almost always has to deal with logically contingent challenges. Intelligence is all about consistently getting those answers right, in a real world occupied by that multiplicity of contingencies, serial complexities, and surprises.

NOW, things can get complicated. When you are trained in all of these fundamental ways, you have the capacity to improve your neurological operations at the highest levels of operation and performance that are especially useful for you, in particular. At these highest levels, your needs will never perfectly jibe with mine, or with any other individual. Those needs might relate, for example, to a history that involves a particular neurological or psychological illness, a history of brain injury, addiction, or infection, or medical treatments that have had impacts on your brain and your life that put you into a special class or are unique to you. They might relate to specific concerns that relate to your employment success, to the effectiveness of your personal life, to those special hobbies and other activities that you cherish, or to that special combination of interests and abilities that sustain your enthusiasm for life. They might relate to the special talents or abilities that you'd like to further improve and sharpen on the path to still-higher performance abilities and still-greater personal achievement. For us scientists and engineers, creating the very elaborate tool sets on BrainHQ that could help you address your special needs, or polish your particular variety of apple and truly excel is a complicated work in progress, because filling out this performance landscape and loading exercises onto the Internet will take us years—and like human ability itself, the brain training help that could be provided is almost limitless. You'll probably have to do much of this work on your own, out in the real world—hopefully guided at least a little by the brain plasticity principles that you've learned about in these pages.

You might be wondering how you could find time in your busy life to actually take on the kind of commitment that would be required to improve or sustain or excel at this wide range of fundamental and higher-level abilities. One of the most remarkable things about your plastic brain is that all of these capabilities can be addressed in practical ways by a combination of your spending a little time each day working with scientifically designed, plasticity-based exercises like those at BrainHQ and by making important adjustments in your everyday life that support and grow your brain health as you live your life following your new brain fitness plan, out in the real world.

Many people ask me if they can expect to get similar benefits by simply making intelligent brain plasticity-guided adjustments in every day life. As I have emphasized earlier, lifestyle changes can carry you a long way in the direction of improving, sustaining, and potentially recovering brain health. Again, they may be all that you need, to retain and to grow your own abilities, from this time forward to the end of your life. At the same time, one advantage of internet-delivered brain training programs is their efficiency for driving positive brain change. That may be especially important for you if you have very good reasons to improve your brain health or to grow your brain powers in ways that get you to a better or stronger place over a reasonable period of time.

To be effective and efficient, we have designed brain training exercises following the following neuroplasticity-science-guided principles:

1. All exercises are designed to rapidly adapt to your performance level. If you're really struggling, they determine that lower level at which you can get the answer right most of the time and work from there to drive you progressively, from your own exercise-determined starting point, in the improving direction. If you're in pretty good neurological shape, programs rapidly adjust to your high performance level, to drive you to a still-higher level from that already lofty position.

2. All exercises are constructed to achieve near-optimal rates of brain change. We now know the brain rules that govern enduring change, and we know how to design brain training exercises that drive changes with the highest possible efficiency. Every exercise cycle counts for driving positive change. We know that we can easily drive changes in ten to fifteen minutes that could take a normal person several days to achieve in their natural life. It's all about efficiently spending your time doing brain exercises.

3. All exercises are presented in a format designed to motivate you. As I have repeatedly pointed out, if brain change doesn't matter for you (your brain), you can't expect to get any lasting value from time spent at brain exercises. Your brain only permits change when it judges that what you are doing is important for it.

4. Every exercise is designed to drive improvements in multiple skill domains. If we had to achieve all of the goals described earlier in this chapter one at a time, you'd have to spend two or three times as long, working on your computer or pad to achieve the same broad benefits. A goal in our refinement of brain gym tools is to continuously increase the benefits for the brain for every minute spent in our web browser brain-training environment.

5. Each exercise is designed to recover, rejuvenate, and empower brain systems—not just individual bits and pieces. We always engage working memory machinery in the brain to guide plasticity from the top down, to make sure that we effectively drive changes at every brain level.

6. The exercises are designed to assure generalization to real-world abilities. The goal is to extend a trainee's behavioral improvements to their natural operations in real life. We have achieved that generalization to real life at the highest levels of exercises, directly confirming that we haven't just trained a person to master a potpourri of brain tricks. After all, if brain training programs don't drive real changes in real everyday lives, they are of little use.

7. Embedded computer-delivered assessment strategies document—and objectively inform you about—the magnitudes of improved neurological and behavioral abilities. By accurately and continuously assessing performance abilities as training progresses, a trainee can know that they are (or are not) changing for the better. Just as you know in the gym that you are getting stronger or faster by the weights you lift, the reps you complete, or the speed that you run, we document a trainee's neurological performance improvements in ways that track brain gains in alertness, strength, flexibility, endurance and speed.

8. The exercises don't waste your time by having you work at things that have no scientifically proven value. Lots of people spend lots of time working on "brain games" that they believe may contribute to their brain fitness, but for which there is little or no scientific evidence that they have any value whatsoever. Unfortunately, crossword puzzles, Sudoku, most Internet-delivered "brain games" and hundreds of other activities argued to be good for your brain fit into this very large category. In our own case, the tools that we have constructed have been demonstrated to drive positive changes in performance abilities that positively impact everyday life in controlled scientific studies, now reported in more than 60 peer-reviewed scientific reports. Many more studies demonstrating program effectiveness have been conducted in-house. A second important class of studies has documented corrective and empowering training-induced changes in the brains of individuals who have used these programs. Our commitment is to limit what we provide in the BrainHQ gym environment to those tools and strategies that have actually been proven to work.

Again, I realize that talking about the virtues of our BrainHQ programs might be interpreted as a selfish act of ascientific commercialism. It is important that you understand how our science has led to the development of effective and efficient training strategies. I want you to think about the values of engaging with a program like this for your own good. I have written this book because, as a scientist and fellow human traveler, I know that these forms of help may be of major assistance for you or for people that you care deeply about. I know that they can help direct many of you down a better path with a stronger and more capable brain, and to a more effective, happier, safer life. If we could provide this help for you for free, and if in doing that, could somehow sustain the further development of what we can provide for all of those millions of people who could benefit from the use of these science-based tools, that would be our price. As it is, access to its use is open, available anywhere in the world, and for what it is, very inexpensive. Objectively, for many of you –and for many other people that you care about or love—spending time using the web-delivered tools at BrainHQ just about the best way that they could be spending their time.[11]

How could I possibly fit brain fitness training into my already extremely busy daily life? The average person spends several thousand hours each year watching television. I'm going to suggest once again that this may be excessive. Consider what you get from this heavy expenditure of time as it relates to your personal growth, and to your sustaining an effective, rich, happy life. If you spent as little as one-hundredth of that time in a well-organized program of brain fitness, your personal returns could be expected to be far greater.

Achievement of any important general goal in training commonly requires that you spend approximately 15-60 minutes per day, three to seven days per week, for a total of four or five to more than one hundred

---

[11] If you visit www.soft-wired.com/ref/ch31, you will find I have provided links to other strategies that have been developed or are applied following these same principles—and have been shown to work in "gold-standard" training outcomes studies. Every commercial provider of 'brain games' or 'brain fitness exercises' makes strong claims about their neuroscientific bases. I strongly urge any consumer of those programs to look behind the marketing to make sure there is not a wizard behind the curtain! Alas, there usually is.

hours at computer-controlled brain training. Training days should be successive, because gains are consolidated in the brain overnight and accumulated gains are stronger when you return to the same task sometime the following day over at least a three-day span in each training week. The total time spent at training can be expected to reflect the level of personal need, and the strength of a trainee's personal determination for self-improvement. If they're really struggling, they'll have to spend more time at a given set of exercises than they would if they had started at a higher initial performance level. If they're really motivated, they can expect to record palpable improvements with as little as one to three weeks of effort. Daily exercises can be completed in separated blocks of time—for example, in a brief session or two in the morning, with another session at work or in the evening. Training programs can be initiated on any Internet-connected computer or iPad, at any time. Because progress is tracked and information about that progress stored in an encrypted (secret, always secure) form in the Internet cloud, a trainee can complete a daily session at home, at work, on a cruise, on an Internet-connected airplane—anywhere that they can hook up via an average-speed Internet connection. You can think of any given major training achievement (like improving your working memory, or revitalizing your brain's learning control machinery, or improving your social panache, or exercising your brains in ways that re-strengthen areas in which Alzheimer's disease first emerges, or working to bring your tinnitus under control, or increasing your flexible intelligence) as involving a commitment that is equivalent to the time that you might spend actually sitting in a chair in a classroom in a course that you might take at your local adult education center or community college.

These crucial changes in the efficiency of your brain operations will likely re-enable substantial and continuously positive personal growth. Your effectiveness in your job and in life should be positively impacted. You can expect to be a more strongly centered person. You should be healthier. You can expect to be more confident. You can expect to be happier. If you're older, years of useful independent living may be saved. There is a significant likelihood that you will live longer. You would likely be a more interesting person, more likely to retain your stature as a fully effective member of your family and of society, less likely to be reduced to a caricature of your younger, former self. There is an excellent chance that you could once again remember more than you forget, and once again operate at speed in ways that make you more competitive with those younger people around you.

The computer or tablet is critical for achieving personal brain health goals in a brain fitness center setting because it enables the maximization of training efficiency. For every minute of training with the precise control that such a device can provide, you can gain the benefits that might be expected to be achieved by spending a large part of each day working on even well designed self-improvement strategies. Computers can monitor your responses, control the finely graded progressive steps that optimize learning rates, easily record a trainee's responding, and provide instant feedback and more complete end-of-training reports documenting gains. We have noted earlier that modern technology is adding to your problems in aging because it is one of those cultural developments that makes it so difficult to keep up, and because, paradoxically, it can contribute to making your brain lazy. At the same time, modern technology has provided us with a set of tools that critically enable the delivery of brain plasticity-based training out into the world. Computers can control your exercises so that they are always driving you to improve, on the cutting edge of your capabilities, giving the brain pretty close to exactly what it needs in each learning cycle to change your brain for the better.

The problem in organizing a brain training program is more complicated for an individual who has a special neurological or psychiatric burden(s). To help individuals with these more serious neurological problems, we have organized special challenges and courses that are designed to guide them through a training path that applies in a more targeted way to their specific history. Challenges and courses are introduced at BrainHQ when scientific studies directly demonstrate that a specific set of exercises has significant and enduring impacts on brain health for individuals in that human cohort. If and when the problem is serious, we recommend that training be monitored and guided by a qualified medical professional or therapist.

If you are incredibly lucky, you may live a life in which you will never need to engage in any web-delivered brain training exercises. I strongly encourage you to try to follow that path. Of course, you probably know someone who doesn't strictly need to go to the gym to work out, but they do it anyway to achieve even higher levels of physical fitness than are possible through everyday activities. If you live your life for the benefit of your brain, you and your brain can be healthy, safe, and full of mental vim and vigor. At the same time, you may want to supplement your brain fitness with brain training exercises to push yourself even further, to achieve your personal goals in a busy life that demands training efficiencies, and to maintain and recover function lost through aging or other issues. It can literally be a life-saver for an individual with significant cognitive decline facing a potentially disastrous older life. BrainHQ is especially valuable for you if you carry special neurological burdens with you that have come unasked for, in your own special passage through life.

A well-maintained and well-trained brain is a very good thing to possess from the perspective of your life and its prospects as you advance in age, and a brain fitness center like BrainHQ is another resource that you can potentially exploit to make sure that you have one.

For further explanations and extensive references and citations
related to the information in this chapter, please visit
www.soft-wired.com/ref/ch31

32

WHAT DOES MIKE DO?

How I Have Organized My Life So I Can Continue to Thrive and Grow

I realize that I'm running the risk of slipping into egomania by telling you how I approach the difficult job of maintaining my own brain fitness. Now, at age 71, I have two main personal goals. First, I want to continue to thrive at an older age. I don't want to waste all of those years that my dear mother lost to infirmity and Alzheimer's toward the end of her life. Like you, I have very good uses lined up for my time on earth, and I just don't want to lose out on them. I want to see my beautiful grandchildren grow to young adulthood. I want to continue to write and work effectively. I want to flourish as a good citizen, husband, father, grandfather, and friend.

But I don't just want to maintain my life—I also want to excel, If possible, and to continue to grow. I want to continue to advance my understanding of things in my worlds. I want to sustain my competitive effectiveness with the (mostly younger) people that I work with. I want to continue to enliven the person that I am. I want to put my brain to the best possible uses as long as is possible.

Science tells us that a key to sustaining and growing our neurological abilities is seriousness of purpose. I am old enough to have retired, but shall not withdraw to a life of ease and comfort because I know that the brain slowly dies when nothing that it does matters to it. If there is a single piece of advice that you carry away from this book, it's to understand that what sustains your brain sustains *you*. You need to continue to work seriously at things that support your brain's health now, and continue to work in ways that support it out to the end of your time on Earth. If you're going to do that in retirement (and many people accomplish that with high success), you need to find new things that are important for you to do almost every day. If retirement means a withdrawal from active life to a life of ease in which nothing matters more to you than your daily dose of passive television viewing or simply relaxing, you are not doing all you can or should do to keep your brain healthy.

I use the Internet-delivered brain exercises at BrainHQ daily, because they were created by me and my colleagues specifically with the goal of keeping aging brains healthy and active, and because I know they work—my own research career has shown that unequivocally. I spend about 20 minutes each day on this form of exercise, and I take it seriously. For me, it provides an easy and efficient way to get a regular dose of healthy brain exercise.

When I am not at work or doing brain fitness exercises, I spend a significant part of every day taking care of my brain in my own free time and in my own ways. I have a small vegetable garden, a berry patch, an orchard, a small vineyard, and flower gardens. I love the look and smell and feel and sound and taste of the things that are out there. I love to find a snake under a garden rock, watch the lizards scurrying to their hidey-holes, follow the ants to their new larder, listen to the flutter of the house wren's wings, or just drink in the sensations coming from my body in the warm summer or cool winter air. I love to watch the bees in their hive, and try to be a good shepherd for the 20 or 30 thousand livestock in my little apian herd. I have fun building things out of clay, wood, stone, or metal; I have a project of some sort under construction on almost any given day. My wife calls me "the farmer and the farmer's wife" because I enjoy preserving the produce from my garden, and making jam, pickles, tomato sauce, and anything else that seems interesting or fun that can be carried from the garden to the table. If I sound a little like a child in these activities, you've captured the spirit of it. My daily mantra is: "See and hear and feel and taste and get your hands back into the world in detail, just in front of you, as if you were still a child."

Every weekend, I do the *New York Times* crossword puzzles. I try to focus to get it done as fast as possible, racing, as it were, against an invisible clock. I do it in pen, and try hard to minimize the number of over-writes by holding large numbers of words in memory. I might add that I know that there is almost no scientific evidence that doing crosswords (or Sudoku or any other puzzles like this) benefit your brain. On the other hand, it is a form of memory test, and so far, it looks like my memory is still holding up.

I complete a jigsaw puzzle with 1,000 to 1,500 pieces once every couple of weeks—usually with some help from my daughters, grandchildren, or visiting friends—because I enjoy it, and unlike those crossword puzzles, I'm quite certain that it's good for my brain. I'm continuously striving to improve my visual neurological capabilities by holding more features in memory and by increasing the number of geometric or color features and pieces in my searching. Mentally manipulating and rotating the pieces gives my brain a workout, too. My wife and I also play cards or games; trying new ones is always a good idea, and grandchildren can almost always be talked into it.

We like to take advantage of the cultural possibilities, large and small, that we find abounding in our city. I'm trying to be continuously alive and closely attentive to the details of the indoors and outdoors around our home and within our outside cultural environs in my external world. The goal is to thrive and excel—to try to improve—in performance and in understanding, at almost anything and everything that is taken on.

I take a brisk 30- to 40-minute walk in my neighborhood every morning that I can. There are approximately 30 intersections within reach during my city walks, so I have found that many different route variations are possible. Still, I find myself working on the mastery of different parts of my neighborhood for stretches of five to ten days, coming at each block from different directions with my attention directed to different details. You'd be surprised by how many new and wondrous things I have discovered out there, close to the home I've lived in for more than 40 years.

I know it's important to try to physically exercise—if I am able to—for at least 30 minutes each day. I know that exercise is good for my body—and at least as importantly, I know that research has shown that it's good for the brain as well. I also know that I can get the most brain benefits from physical exercise if I don't shut down my brain while I'm exercising. I might prefer listening to music or letting my thoughts drift while I walk, but from a brain health perspective, that's a wasted opportunity.

My strategy is to pay close primary attention, in different parts of my walk, to two of my brain's most important worlds. About two-thirds of the time I try to reconstruct everything that I'm hearing, seeing, and smelling in the external world, continually looking, listening for, and smelling out those many little visual, auditory, and olfactory landmarks and surprises that are a part of every trip that are out there for any attentive walker. The brain loves those surprises. Several times later each day—and as a habit, immediately after I return from my walk, and again when I first get home from work—I reconstruct that morning's adventure in my mind, in as full and vivid detail as possible.

A longer-term goal is to become the master of my local environment, by which I can remember and completely mentally reconstruct every detail in the part of the world that I operate in. I know that my human ancestors in their natural environments all accomplished this. I know that when I do this, I am exercising key faculties that are important for me to retain, to protect my brain from a decline into end-of-life trouble. While the navigation-related training programs at BrainHQ are designed to recover and strengthen these abilities, I am pretty certain that this real world training is also important for growing and then retaining them in tip-top shape.

In roughly the other third of my walk, I pay attention to my corporeal world. I feel my body and sustain a lively conscious image of my body and limb positions in three-dimensional space in detail as I move. I try to move with full-body coordination, remembering that I have soles on my feet, a neck, arms, legs, and—most of all—shoulders, hips, and a spine. I change the speed, length of stride, and posture in walking to add variety to my pace and exercise my brain's ability to make all of the necessary adjustments, as I introduce those variations. This is especially useful when I can walk off-pavement on a mountain path or across hill and dale in a park. I have invested in shoes that help me feel the ground beneath my feet better. I don't want every step I take to be predictable. I am sure that there are other adults and quite a few children who think that I'm a little comical as I play out these walking shenanigans in my neighborhood. I don't care. I think that in a few years, I'm going to have lots of companions out there working hard to retain and grow their neurological competencies.

A couple times each week, I substitute another form of exercise for the morning walk. Whether it's in the form of vigorous work in my garden or shop, an enthusiastic bout of table tennis, or a brisk round of golf, I know it will be just as good at elevating the blood flow and turning up the dimmer switch in my brain, but with the added benefit of supplementing the physical activity with higher cognitive effort. When I'm thinking forward to the next golf hole or shot, or when I'm actively guessing what my ping pong opponent is going to try to trick me with, or plotting out my garden's future or remembering its past, I am strongly exercising those higher areas in my brain that I want to keep in good shape if I'm to keep myself safe from later unpleasant neurodegenerative news. In all of these endeavors, it is important that I try hard to improve in strategic planning and rehearsal and in fast responding, coordination, and accuracy.

My wife and I work hard to bring social-activity exercise—and the joy that positive human interaction can bring—into our lives. When I walk and meet someone else on my path, or as I operate out in the wider world, I am determined to bring as many positive moments in the lives of others as I can. I know—if the message is properly received—that this is good for their brains, but I also know that it is good for my brain. Delivering little doses of happiness out into the world makes me happy. If I want to keep my dopamine machinery alive, this is a very good and simple way to do it.

We also try to find lots of ways to extend our help and friendliness in as many good directions as we can. I know that this provides just a little brain health assistance to our family and friends, and helps us in the bargain. We try to turn what could be lonely occasions into social events—whether it's inviting people to help make wine, olive oil, jam, or a birdhouse, or to play a game of bridge, look for mushrooms, pick the blackberries, or walk over the ridge. When friends join you for any good purpose, you all exercise the brain's social cognition and happiness machinery! Besides, you get good help and can be the recipient of lots of good spirits. There really doesn't have to be much of a reason to find a way to turn a lonely egocentric day into a social extrocentric day. If you want to resist those powerful forces that seem to want to turn you into a self-oriented caricature of your former self, this is an important piece of that puzzle.

I read two or more books a week through the year, half for fun and half for understanding. I read dozens of scientific reports each week. I listen to the news on national and international public radio almost every morning as I am beginning my day. I scan a newspaper (or news online) almost every day. I want to continue to grow my understanding of the wider world. I seriously seek to grow in understanding and wisdom.

Finally, I try not to do things that will drive me in a negative direction, and I try not to pass up opportunities that grow my ability to operate independently in the world. I try not to watch too much television, and turn it off altogether on the weekends; PBS documentaries, old Western movies, and baseball games are my principal vices. I try not to spend all day sitting in a chair or on a sofa. I don't have a GPS device. I make extensive use of the wonderful modern tools that technology has provided for us, and that so clearly enrich our lives—but I consciously try to keep them in their proper place as tools, and not as a substitute for my brain or my life. This is not always easy, of course, but I try!

My wife and I don't go on vacations that are planned or guided by others. I'm now determined that you can't make me drive somewhere when I can easily walk there. I am generally not afraid to take a chance. Surprises, problems to solve, unexpected adventures, things not quite working out as expected—from such occasions come the experiences that enliven a life, and add a few of those stories that others might actually want to hear.

This little egocentric excursion is not meant to be a paean to "Marvelous Mike." I don't live in a perfectly brain health-oriented manner every hour of every day—I'm human, after all. But I do try to practice what I preach. Your life is richly different from mine, and probably the better for it! I've been fortunate enough to have lots of material advantages—and advantages of good health—that we may or may not share. This personal exposé is simply meant to provide you an admittedly sketchy human example of what it means to think about organizing your everyday life around brain training. There are a million other things that you might do, in simple daily adjustments, on the path of personal growth or recovery. If you effectively apply lifestyle changes—perhaps also with a regular schedule of visits to a brain fitness center—you'll very substantially increase the chances of living the last decades of your life in good spirits, and in very good brain health.

For further explanations and extensive references and citations related to the information in this chapter, please visit
www.soft-wired.com/ref/32

33

CAN'T I JUST TAKE A PILL?

Other Useful Strategies that Can Contribute to Your Brain Health

Everyone who wants to maintain body and brain health, and who is capable of physical exercise, should try to make time for a significant daily dose of it. I noted earlier that research has shown that doing at least 30 minutes of physical exercise each day confers substantial neurological benefits. Physical exercise increases cerebral blood flow, and who wouldn't want more oxygen delivered to their brain? It also stimulates the release of those chemicals that are released from your "dimmer switch," and who wouldn't want to have the lights turned up just a little higher? And, hey, I don't have to tell you that it's good for your body, too!

I emphasized earlier that you are losing a major opportunity for improving your brain health if you are exercising your body in ways that disengage your brain from participation. Working on exercise machines in a (brainless) fitness center is an example of how not to make the best use of your physical exercise time. Machines that eschew variety in movement control, while promoting stereotypic movement—that great enemy of healthy and flexible neurological control—should be particularly avoided. You would never exercise your brain by solving the same problem a million times using exactly the same approach to come up with exactly the same answer—but from the brain's movement-control perspective, that's exactly what you're doing when you spend all of those hours working out on many of those exercise machines.

Formal exercise gym environments also commonly isolate movements to parts of your body—which would be just great if your body were designed to operate as a composite of disconnected mini-robots. It isn't. In your natural behaviors, the core of your body is almost always in play. Isolating movements of an arm or neck or leg without a contribution from the hips, shoulders, and spine is highly unnatural.

Most individuals who work out in a fitness center also pay only limited attention to the feelings of the movement, or to the visually or kinesthetically monitored flow and perfection of them. Movement control is all about your brain interpreting those immediate feelings along with those beautiful products of sensory feedback—translated so elegantly by your vision and your kinesthesia—coming back from your body. "Senseless" exercising is good for your strength and vitality and is known to help get more blood and oxygen to the brain; still, for exercising your brain as the controller of your movements, it's largely a waste.

If your main dose of exercise comes from a daily walk, run, or bike ride, it is important that you also spend some time in those exercise periods paying attention to those feelings. When I see a walker or runner or biker disconnected from their physical activity or the world they are moving across, pounding their legs in stereotypic movement while they listen to their headsets or stare blankly ahead, I see someone who is not making the best use of that precious daily physical exercise period.

If you are a daily walker or runner, you might also think about how you can elaborate the control of all of those movements that are not effectively engaged in your regular daily exercise schedule. I have recently adopted the strategy of spending about five minutes each day working on the elaboration and control of a specific class of movements—for example, the control of my head, or hips, or jaw, or left knee, or left thumb, and so forth. I work for four or five days in succession, focusing on whole-body engagement but with my attention always centered on movement around the target joint(s), trying my best to elaborate and refine control for all possible trajectories on the path to goal achievement, and to improve my speed control and the fluidity of my motions.

The value of walking or running as a daily exercise stratagem in an urban environment is limited by the paved surfaces that we are compelled to move across. Moreover, just about the only way that our feet can tolerate all of that pounding on these perfectly smooth, flat-rock surfaces is to move across them in hard-soled shoes. The clear value of just taking those shoes off is offset by the difficulty one can have in finding any natural surface to walk or run across. Still, if you find yourself in an environment in which the opportunity avails itself to actually feel the ground beneath your feet when you walk or run, then give it a try. The beach, meadow, pasture, or forest path are all good places to find those natural surfaces that your feet were actually designed to trod upon. Once your lightly clad or bare feet are toughened up by such exercise, your brain will thank you for providing it with the endless challenges in fine motor and balance control and adjustments to small perturbations in vision (what scientists call "visual slip") that the natural physical world so richly delivers.

There are six simple brain plasticity-based rules that you might think about following in planning your own physical exercise regimes:

1. As you move, focus on the feeling of the flow of that movement. Work hard to progressively improve that flow, and the achievement of your imagined movement targets.

2. Move with your whole body. You have a flexible core and a spine. Use them.

3. Rigorously avoid stereotypic movements. If my goal were to extend my reach to grasp an object at some point in space, there would be limited value in practicing that specific movement task a million times. From my brain's perspective, it would be far more useful to do my best to reach one hundred nearby targets via one hundred different paths at one hundred different speeds—al in the end, under my progressively more perfect control.

4. Include postural variations and weights. These variants are important contributors for adding richness and variability to your movement-exercise programs. Keep in mind that for different movement learning targets, richly varying the challenges that you present to your brain and body is a key.

5. Monitor the quality and precision of your movement. Reward yourself (in your mind) for every little improvement. Try as hard as you can to achieve perfection.

6. Set as a goal the mastery of all movements at a wide range of possible speeds. On the one hand, you want to perform your movements at a fast pace when appropriate. On the other hand, seek perfection for controlling slower forms of these same movements. The word for that is "grace."

There are many other strategies for exercising your body in ways that work for you. For example, my wife participates in an aerobic exercise class that continually changes their exercise routines. That's exactly how such activities should be constructed. Many other citizens engage in other regimens of continuously challenging, never-masterable physical exercise. Tai chi is an excellent example. Squash, tennis, and ping pong are progressive physical activities that can work for you if (and only if) you take improvement and skill elaboration at them seriously. Ballroom dancing provides another good example of an exercise form that can be wonderful for body and brain, if you are dedicated to developing your skills and abilities in the ultimate direction of—but of course never fully achieving—performance mastery. Bicycling is especially valuable for you when you vary your travel routes, and when you dare to ride across unpaved terrain. Whatever activities you choose, remember that regular—ideally, daily—exercise is an important goal.

Many people now understand that a diet that pays some attention to brain nutrition is also a key to good health. Like all body tissues, your brain needs fuel and essential nutrients. It has some important, quite specific needs. There are excellent guides to healthy eating, as it applies specifically to your brain.[12] At the same time, remember that the continuous remodeling of the detailed functional circuitry of your brain required to improve or sustain its functionality can't be delivered to you by mouth! Food is like the gas in your car. You certainly do have to gas up, keep your radiator full of water, fill the oil tank, and make sure you get a few key additives in there. For you, this means eating fruits and vegetables of all colors, not passing up the cold water fish, consciously limiting the consumption of fat, refined sugar and processed foods, and keeping yourself well hydrated. Of course, the harder and more sophisticated job is to keep that car engine and all of those moving parts of the vehicle in good repair—which is what brain fitness is really all about. And you can't get it at the dinner table.

Your ability to sustain your mobility and independence and brain health are of course endangered by overeating. Obesity and incipient Type II diabetes are bad for brains and foretell a shorter brain span and life span. Steering clear of too much of the wrong foods is one of the most important steps for many individuals on the path to improving their brain health. I've personally found great help in the diet-control mobile apps that help me calorie count and monitor daily fat and

---

[12] See www.PositScience.com for specific recommendations.

carbohydrate consumption. I've also adopted a strategy documented to be neurologically beneficial (and in other ways healthful) by scientists at the U.S. National Institutes of Health, in which normal eating days alternate regularly with light-eating days.

Most older individuals are also feeding their bodies with a rich supply of prescription medicines and supplements. Americans over the age of 65 have an average of about 40 prescriptions filled at the drug store each year—at a cost of about $3,000 per person. The average 65-year-old regularly takes three prescription medicines; by age 75, ambulatory individuals are taking a rather astounding average of nearly eight drugs on a regular basis. A large proportion of this medicine targets the vicissitudes of old age. Of course, pharmaceutical treatments are hardly limited to older populations. In modern medicine, drugs represent the primary treatment for almost every human condition.

Prescription medicines are essential for maintaining the health of our citizens, and they save many adults among us from dysfunction, distress, and even death. That prescription medicine that you may be taking may very well provide you with important additional years of healthy life or sanity. At the same time, there is also very little question that modern medicine is drug-happy. The appeal of solving every problem in our lives by taking a pill has deeply infected our general population. Many people, faced with their own decline, are waiting in their easy chairs for a medical breakthrough—a new miracle drug—to save them. "As soon as they develop that new pill," they tell themselves, "Alzheimer's disease will be a thing of the past... hopefully, just before I get it!" Or perhaps they might tell themselves "Now that I've started to get the shakes, I wish they'd hurry up with those stem cells. I might need some pretty soon now!"

You've begun to understand, I hope, that your brain is just not that simple. Your functional abilities at any point of time are expressing the unbelievably complex state of wiring of your plastic brain. The only real way that a person can recover memory or other mental or action control abilities is through the engagement of brain plasticity processes that progressively drive those many degraded faculties in corrective directions. You have to plastically rewire your brain to correct these problems. Drugs can often help, but they can never be the complete or even the primary solution for a substantially struggling brain. Your brain has to learn its way out of this kind of trouble. Fortunately, you have a richly plastic brain to help you get better and stronger—and if need be, to help you claw your way back out of that hole.

I don't mean to say that drugs aren't life-changing and at times, indeed miraculous. You or someone you care about most likely takes one or more drugs for a neurological burden or other health problem, and in the majority of cases, these drugs are absolutely essential for health and well-being. At the same time, remember that your brain is plastic. For neurological issues, there is almost always more we can do to help ourselves beyond opening that medicine bottle every morning. We can use our brain plasticity along with appropriately prescribed medications to do a more complete job of neurological restoration.

Just as it is important to understand that a special neurological burden does not forestall your making good use of your natural brain plasticity, so too is it important to understand that you don't lose this gift just because you suffer an impairment in hearing, vision, balance, or movement. For an individual carrying one or more of these common forms of neurological burden, your goal should be to modify your brain training focus to exercise those resources that you can still dependably engage, while not forgetting to work on those abilities that you are losing to get the most possible out of them. Remember, always, that mental exercises have substantial neurological value. Even if I am not able to get up off my chair and dance, I can mentally rehearse that dancing as I listen to the music, or can mentally rehearse playing the piano as the concerto goes by in my mind. And when it is difficult for a person to support their weight on a frail or injured or unresponsive body, they can give it a boost in overcoming those powerful forces of gravity by exercising on a mat on the floor, or in a swimming pool. All kinds of movement control and practice may be possible when we take advantage of the floor or buoyant water to take some of the weight off! Perhaps some of that recovered strength and control can be brought back out to good use, when you get up from the floor or come back out of the pool. If we are serious and systematic enough about it, all of us can find many ways within our own achievable performance repertoires that can contribute to more effectively exercising our brains to sustain and grow its fitness.

Every one of us has been given the great gift of brain plasticity. It provides us with a completely natural source of help that can drive improvements or corrections for us, potentially on a large scale. These processes are fundamental to our living brains. Our brains don't care if we have complicating factors coming from our psychiatric or neurological history, or if we have a long history of taking medicines for dealing with our personal and perhaps long-standing problems. Brain plasticity is there for us, ready to engage as a resource to assist us in our recovery from injury, infection, or disease. It is there for us, every day of our lives, to help us grow in strength and ability, and to increase our resilience and our effectiveness. Those pills that we take—or all of the brain healthy foods or supplements that we consume—are helpful in maintaining the health of our brains, but they are not capable of truly restoring our neurological capabilities. Our brain plasticity must play a central role in any true recovery, and in any very significant behavioral performance improvement. Exercise right, eat right, and take the medicines that your doctor prescribes, of course. But also, act right, by reorganizing your life with your special brain fitness needs in mind.

For further explanations and extensive references and citations related to the information in this chapter, please visit www.soft-wired.com/ref/ch33

34

MY MEMORY IS PRETTY GOOD,
BUT MY BACK IS DRIVING ME CRAZY...

Helping Your Brain Help Your Body

In many chronic clinical conditions, and as we age, many autonomic processes—like those that keep the food moving down the throat and through the gut, keep those important sphincters clamped shut, control adjustments in blood flow accommodating changing postural or body-use circumstances, and so forth—may slowly fail us. If our goal is to strive for a healthier, safer life, it is important that we give some thought to how we might address physical problems beyond movement, by asking our plastic brain if it has any help to offer. You will probably not be surprised to hear that it does.

About one in five Americans lives with chronic pain. Not surprisingly, once you look at older people, the numbers rise to one in three, causing even more discomfort and agony in the later decades of life. Whatever the age, it is challenging to remain happy and productive when you hurt all the time. Because chronic pain often limits physical activity, it can also limit strategies available for strengthening or recovering broader neurological capabilities. It would be a real bonus if brain plasticity had something to contribute to helping an individual who has to live with aches and pains. Again, it does.

Scientific studies have shown that the loss of mobility and/or the loss of a person's driver's license foretell a premature end of personal independence. It would be terrific if the brain of such an individual could help them sustain their walking and running and safe driving until they come closer to the very end of life. If they're lucky, it probably can.

Of course, while chronic pain, autonomic difficulties, and loss of mobility are more common in older adults, they can happen to anyone, for any number of reasons, at any age. Bouts of chronic pain are a part of virtually every person's journey through life. Problems with mobility can arise from a thousand different causes. Before I discuss how we might think about putting up a good fight and knocking a few of those aches and pain down a peg or two, or potentially forestalling that loss of personal or vehicular mobility by implementing a brain plasticity-based approach, I think that it is interesting to reflect on how these problems have grown in modern societies.

Forty years ago, about 5% of American women who underwent a radical mastectomy suffered from chronic pain. Today, 50% do. Medical scientists are masters at trying to explain away this kind of epidemiological difference. For the most part, you women may be interested to hear that they attribute it to growing female whininess. These experts just don't know the women that I live and work with! What could possibly account for it?

When you remove a richly innervated female breast, the brain's representation of the skin of the body has to remodel itself on a large scale, to re-occupy the large territory formerly representing the now-missing breast with highly locally coordinated sensory information from the intact skin. The re-establishment of strong non-pain inputs again fully engaging this radically altered brain area is, by its nature, anti-pain. The remodeling of the woman's brain map of the body surface is dependent upon the rich, natural stimulation to the still-intact surfaces of the body. Modern women–the great majority of whom now spend much of every day in a semi-sedentary workplace, or at work or at home sitting in front of screens–often just don't move very much. Sitting in a chair for hours at a time with a largely immobile trunk is not exactly what the brain requires to deliver an adequate dose of anti-pain brain reorganization. Get thee in motion, dear lady, if you have had to endure a surgical procedure like this! Teach your brain not to hurt, by assuring that those non-painful inputs that help you re-engage those brain areas that used to respond to your breast have good, new, anti-pain work to do. It won't reduce that pain in every case, but at the very least it should be a measurable step in the right anti-pain direction.

To cite a second example, 40 years ago, knee and hip replacement surgery were in their infancy, and only a handful of patients had undergone these procedures—a large proportion of them, with marginal outcomes. In the present day, each year there are well over a million individuals driven to their doctors because of their ongoing unendurable discomfort or pain to have hips and knees replaced in the U.S. alone—at a cost of more than $13-14 billion annually. At least half that many other people show up every year in different surgical theatres, suffering beyond endurance, to voluntarily endure major surgery on their spines. In our modern world, almost every town and city neighborhood has a "pain clinic" that is an active center of drug dispensing, physical therapy, and surgery, all dedicated to dealing with the suffering that plagues so many of us in our modern societies. Even given an incredible (and incredibly costly) effort to combat it, the prevalence of chronic pain in elder populations is still growing. Almost no one asks the obvious question: "WHY?" Again, the usual answer from the medical science side: "Modern citizens are just wimps."

Don't take that explanation too seriously. Think of yourself, or the people around you. Don't you know when pain is unendurable? Are you a wimp?

As you are reading these words, I would guess that you are sitting down. Do a little experiment for me. Set your stopwatch tomorrow, and record all of those times through your day when you are sitting, with your knees, back, hips and shoulders out of action. Don't forget the time spent in your car, at your desk, in your chair or couch staring at your computer monitor or TV, or reading the last chapters of this book. If you are a typical modern citizen, you may be a little bit shocked by that accumulated time you spend with your knees, hips, and backs in an inactive mode. How can we expect to adequately support our sustained mobility, when, since childhood, we've been spending the majority of our waking hours sitting on our backsides?

You might intuitively guess that the critical perfusion of these tissues that supports the lubrication of your joints and the immune response in these tissues is not very strongly engaged when those joints and back are so infrequently put into action. In addition, because of disuse, the sensory feedback that contributes to that regulation of blood flow and anti-pain control is grossly neglected, and thereby degraded.

You may have experienced an injury that was attributed to exercising too much, so when I say that pain may be caused by a lack of exercise, it may sound counter-intuitive. What matters is the form the exercise takes. If I were to exercise in ways that stereotypically engage my knees or hips under conditions in which I never monitor my feelings or performance abilities and do not introduce variety into my movement repertoires, I would gain all of the disadvantages of the thousands of moments of micro-trauma that can come for each pounding step, and none of the advantages that can come from refining my neurological feedback power in ways that increase knee and hip anti-micro-trauma perfusion—or that re-establish more versatile, more powerful, and more variable joint control. The wrong kind of exercise can be destructive.

The right kind can hypothetically contribute greatly to your resilience and your pain-free joint use.

Before I move on to the next topic, I want you to remember that Mother Nature has endowed you with a pain sense to alert you to the fact that whatever hurts has undergone damage. As a general rule, you should not ignore these warning signs, because they might be telling you about something that needs medical clarification—and very possibly, serious medical attention. By all means, you should see a doctor. Just remember, while you're seeking that help, that pain is in the brain and mind, as well as in the body. And just as with your other senses and all of your other abilities, its neurological powers can often wax, or wane, through the operations of your brain's plasticity processes.

There are several other important things to remember as you think about how you might engage your brain to help you be a better steward for your body:

1. Most importantly, seek as many reasons as possible NOT to sit down passively. It's addictive, I know. But it's key that you try to limit it. This seemingly small change is related to the health of all of your brain's movement control operations.

2. Keep in mind that positive brain plasticity requires that you pay attention to the feelings of your movements, and to your visually and kinesthetically monitored movement outcomes. This control can work for you if there is rich variety in your movements. It can work against you, if there is little or no variety.

3. Consciously remind yourself that you have a whole body made up of parts that work best when they work together. Only robots are designed to control movements around one or two joints at a time. When we move our fingers, our natural movements should have us gliding our hands, moving our wrists, elbows, shoulders, spine, hips—moving it all. Focus on re-discovering your hips, spine, and shoulders and involving them in movements.

4. Remember that a strong, refined, detailed, and coordinated representation of information from any given region of your body is, by its fundamental nature, anti-pain. When you keep your corporeal world in tip-top shape, you're doing a lot to reduce the probability that you will spend a substantial part of your life in torment.

5. Your pain is undeniably real—but, at the same time, reflect on how the pain is affected by your brain, in the sense that the context of your hurting can contribute to how much you do or do not suffer from it. I do not mean, of course, that the degeneration of your spine or the breakdown of your knee is not real, or is not a constant source of pain. I know how much that can really hurt and have experienced it myself. The subtle difference is that the magnitude of the suffering is contributed to by factors that aren't strictly attributable to that pain-sense input coming into your brain from your body. Try hard to put your pain where it belongs, back into the deep dark hole that it sprang from. This is, of course, not always possible. But at the same time, remember that if circumstances were just right—for example, if you had just won a big jackpot in a gambling casino—you know that your pain might be forgotten, at least for that moment. Try to achieve that moment of forgetting in your mind, to the extent that you can without actually winning that jackpot. My own strategy, when I ache or hurt, is to focus, as much as possible, on the non-pain that is almost always hiding behind the pain itself.

A consideration of the many other ways that your brain influences aspects of your body health is so vast that I would have to write another whole book. Suffice it to say that your brain is playing a significant if not central role when you can't help yourself from eating too much, you stumble and fall, lose control of your bladder or bowel, struggle to swallow, continually drop your silverware, hurt every time you take a step, have a gut ache, have sore gums, find that your food always tastes metallic, get a headache every day at about the same time, can't seem to get a good night's sleep, suffer from that ringing in your ears—and hundreds of other vicissitudes like these. If any such thing impacts you, remember, always, that the locus of control of all of the elemental faculties that control these processes resides at least partly within your skull and spinal cord. You learned to control most of these processes, through brain change, in earlier life. Your learning machinery is still alive, and remains a resource that you can potentially call on to provide at least a little help. For any of these abilities, if brain plasticity is deployed in earnest, it may be possible to be just a little bit more in control tomorrow then it was today. If a tiny average gain were to be achieved each day from this day forward, you may be in a much different place with your problem a year from now. While medicine may also be critically necessary—as it often appears to be for this class of problem—don't forget that you possess a second, potentially powerful asset to call on for help: a plastic brain.

For further explanations and extensive references and citations
related to the information in this chapter, please visit
www.soft-wired.com/ref/ch34

35

DRIVING MATTERS

Navigating the Modern World

By your 60th birthday, there is almost a 50/50 chance that your field of view has become significantly smaller, and an even higher probability that you could be easily shown to be substantially slower in recognizing or responding to danger. These and other changes affect driving abilities and ultimately threaten a person's ability to sustain their independence at an older age. For a rich variety of reasons, problems of this nature can also impact an individual of almost an age.

These growing neurological weaknesses put a person at significantly higher risk for having a serious driving accident. By the age of 80—unless we do something about it—the majority of us represent a real source of driving danger to ourselves and to everyone else on the road. If our driving abilities are left to deteriorate, they must ultimately result in the loss of the right to drive; someone is going to be justified in asking that you give up that driver's license.

But giving up your license to drive causes many more problems than merely losing your personal driving freedom. Scientists have shown that in our modern car-dominated culture, individuals who give up or lose their license are four times as likely to die over the next year, when all other differences between individuals are accounted for. When it comes to longevity in the modern era, it is statistically more important to maintain your mechanical mobility than your personal physical mobility!

Why don't more people acknowledge that they are less than perfect drivers at any age—but especially as they age? Whatever the age, how many near-misses or fender-benders does a person need to experience before it dawns on them that something should be done about their limitations behind the wheel? It's pretty important to accurately self-assess your abilities because the very good news is that, with very few exceptions, a less-than-perfect driver can restore their accurate, wide-field, fast-responding, fast-acting vision in ways that make them safer again, with just a few hours of brain training.[13] The benefits of such training obviously also greatly amplify and extend the way that you drink in information from that visual world when you step out of that car.

For further explanations and extensive references and citations related to the information in this chapter, please visit www.soft-wired.com/ref/ch35

---

[13] For example, see the "Driving with Confidence" challenge at www.BrainHQ.com.

36

IMPROVING YOURSELF

Taking a Holistic Approach to Improving Your Life

My friend Paul is a Chinese immigrant who purchased a small farmstead for his father and mother to live on in their retirement. Through an older neighbor, he met an elderly gentleman, nearing 80 years of age, who encouraged Paul and his father to plant pinot noir grapes on their land. This new acquaintance, Louis Martini, was a famous Napa Valley viticulturist. In the end, the neighbor and Mr. Martini voluntarily made cuttings from their best vines in their own vineyards to graft onto rootstock to help establish Paul's father's vineyard. These three older men, not far from the end of life, shared a great new adventure with Paul and his family that has been the source of countless happy events and stories. More than 20 years later, their vineyard is world-class, with most of its grapes used to produce a super-premium Napa Valley wine. Paul has made a wonderful transformation from nuclear physics to agronomics. His family richly exploits their country life to the benefit of their family and friends. This wonderful personal growth has been extended to dozens of other people—all a beautiful product of the friendship and collaboration of three adventurous older men. Their high spirits led to hundreds of new adventures, and served as the impetus for developing many new skills and abilities for everyone in Paul's kinship.

You are a remarkable composite of a billion or two self-referenced memories. It may be important for you to re-invent yourself at least a little, to change your lifestyle for the benefit of your brain, and possibly, to spend some time doing brain training. The more that you succeed at improving your brain's ability to record and use new information, the better you will be able to sustain and grow that person that you are, in its full, rich, elaborate, and fascinating detail.

At the same time, a vigorous brain health approach also necessitates that you continuously expose yourself to useful new things to remember. In this book, I have tried to help you understand that brain fitness is not going to be achieved just by going to school again, reading good books, or gadding about town. Such activities cannot, by themselves, increase your memory, grow your intelligence, restore your brain health, or reliably protect you from ruin. On the other hand, growing the person that you are also requires exactly that kind of conventional learning. Once you've increased your brain's power, you can much more strongly exploit all of those benefits that you can gain from new experiences and information-gathering.

In thinking about your brain fitness as it relates to re-growing you, consider the following examples—remembering, of course, that you're special, and that your best ways to re-invigorate that personal growth shall necessarily be all your own:

1. Initiate some new (large or small) activity, almost every day. There are hundreds of small and large new adventures awaiting, in your home, garden, neighborhood, community, countryside, the world. Don't let a day pass without at least one new memorable experience, to add just a little bit more to life.

2. Seek and take delight in the positive surprise as much as possible. If you really do believe that you've "seen it all," I'm a little worried about you. Life is absolutely loaded with positive little surprises, if you're actively looking for them. Watch a small child just exploring their environment, and you can plainly see what a delight those surprises once were for you. Be a little bit of a bright-eyed child again. Live more in the wide-awake present. Embrace and be amused by the peculiarities and the wonders and the downright astonishing things that you can't hardly avoid tripping over, almost anywhere you go in your world.

3. Re-grow your savoir faire (the confident self). On page 324 in my edition of Mastering the Art of French Cooking, Julia Child describes a nice daube de boeuf á la Provençale that I can recommend. Why don't you invite someone that you are fond of, or would like to be fonder of, over for dinner tomorrow? As you adopt your own personal strategies for exercising your brain, or as you log out of the brain fitness center, you will feel your confidence returning and you will know that a bolder

247

you is fully capable of taking on that new challenge.

4. Go out of your way to make new friends. Each new friend you make is incorporated into you. Your soul gets to borrow a little bit from each of them, and is always the more interesting because of those golden gifts. Remember that you are also delivering an equally valuable gift to their souls.

5. Aggressively use your sense of happiness and pleasure to help you identify the kind of activities that will most effectively add to "you." Have you ever noticed that you tend to laugh the loudest and most joyfully when you are with other people? It almost never occurs unless you are really intensely engaged in conversation, are in other close personal interactions, or are out in the world with others observing how hilarious that world can be, first-hand. Laughter is one of the best medicines, and laughing and spine-tingling and even quiet pleasures can be good general indicators of those kinds of things that are positively contributing to a growing you.

6. Go to a cultural or social event that you always thought you didn't like, and try to figure out why all those other folks like it so much. We are creatures of our culture that, within our short lives, have knowledge of and sample just a very small part of what other people amongst us have learned to love. I have a friend who is a devoted bird-watcher. Her friends include other devoted bird-watchers. They're not dumb or boring people. Maybe there's something to bird watching that you would enjoy. Or, I'll bet, to a thousand-and-one other things that would make you more interested and interesting.

7. Re-define yourself, if you see clear bases for improving that person looking back at you in the mirror. It is never too late to think about improving your humanity and elaborating and enriching your capabilities. If you realize, upon reflection, that you have allowed yourself to simplify into a form that has lost its earlier subtleties and flexibility, and are now operating in a version that is harder to like, a little work on a personal course of correction may be in order.

8. Monitor and assess yourself. Most individuals slip unaware into that narrow silo that defines them as a caricature of their former selves. Develop good habits of self-assessment, and pay attention to other people's reactions to you that might inform you about how you have begun to move backward in life. And when you give yourself a "How am I doing?" score each day, take the next moment to remind yourself that it could be a little higher for the rest of the day—and at least a little higher still, all day and tomorrow!

9. Continuously enrich your life on small and larger scales. I said at the beginning of this book that we are not limited in the repertoires of our behavioral abilities like one-trick ponies, or always sideways-walking crabs. As a veritable endowment of our human natures, the world is our oyster. You shall be richly rewarded for expanding your little part of the universe and growing yourself.

For further explanations and extensive references and citations
related to the information in this chapter, please visit
www.soft-wired.com/ref/ch36

DR. MICHAEL MERZENICH, PHD

TODAY IS THE FIRST DAY
OF THE REST OF YOUR LIFE

Begin the Transformation to a New,
Better Place in Life Right Now

"Neither a wise man nor a brave man lies down on the tracks of history to wait for the train of the future to run over him." –Dwight D. Eisenhower

"I am the master of my fate; I am the captain of my soul." -William Ernest Henley

This book is a gift to you from brain science. How much you take it to heart, and use it as a reference to rejuvenate and to sustain you, is now in your hands. Participation is voluntary. I hope that you now understand that if you opt for it, brain fitness will require substantial, regular efforts and perhaps some lifestyle changes on your part.

If you are younger, you may want to establish a higher performance level and sustain your abilities through life. Your brain can carry you to higher levels of ability, while increasing self-confidence and positive good spirits. As the years pass by, if you can grow and sustain these critical personal resources, you should be able to grow your integrative and analytic capabilities. With your primary faculties more intact, you can sustain your competence, leadership, and productivity up to that day that you choose to retire. And if (or when) you choose to retire, you can use your newfound understanding of the benefits of life-long brain fitness training so you can understand how to retire without delivering your brain and body to the trash heap.

If you have been dealt a challenging neurological card at any age, remember, always, that your brain has powerful resources to help you gain and recover ability, and to establish, recover, and grow better control of its actions. Brain plasticity is the stuff of life. As long as you're alive, it's with you as a precious, exploitable asset. Don't neglect to take full advantage of it.

We all know someone who's been waiting for a "memory pill" or an "Alzheimer's pill" to save them or keep them safe. Many people believe that eating fish, blueberries, or kale will make the brain healthier. Some folks see a daily bout of physical exercise as the magical elixir, while others feel that a daily crossword puzzle is the key to staying sharp. Meanwhile, if the brain could talk, it would be saying, "These things are helpful, and I'm glad that you're doing them, but I could sure make use of some *real* exercise!"

If you choose the course of brain fitness, it's important to face up to the efforts that may be initially involved. Depending on your age and lifestyle to date, you may have spent awhile on the downslope already, neglecting your brain. In that event, recovery and maintenance will require a serious effort on your part. Your program of brain fitness will require that you quickly shed what might already be well-established negative behaviors that are slowly dragging you down. They may also require that you re-establish a bona fide seriousness of purpose in life—perhaps for the first time in a long time. From the platform of a "life of ease," a real change in the trajectory of your life may be exhausting to even think about!

But the payoffs can be enormous. They go far beyond remembering the odd face or the word on the tip of your tongue. Brain fitness is about retaining your vitality, your zest for life, your independence, yourself. It is all about giving your brain an excellent opportunity to last as long as your physical body. It is about living longer, alive, full of it, fun, still intense, still confident, independent, still growing, more capable and more interesting next week and next year.

Natural forces are working against this fresher, healthier attitude. It's easy to slip unawares into a caricature of your former, more elaborate, and more interesting younger self. As you age, the decline of the machinery that controls your learning and memory also controls your verve, your brightness, your confidence, and your spark. It's often difficult to rise to the challenges of revitalizing your life if your dimmer switch is turned down so low! Be encouraged to know that if you just give it a chance, you'll be surprised how much the joy, the spirit, the dancing of body and mind, the bright daylight of a sunshine-filled life can come back for you.

And your brain will thank you for it.

For further explanations and extensive references and citations
related to the information in this chapter, please visit
www.soft-wired.com/ref/ch37

ABOUT THE AUTHOR

Dr. Michael Merzenich, PhD is a Professor Emeritus at the University of California at San Francisco, a member of both the National Academy of Science and the Institute of Medicine, and the co-founder of Scientific Learning and Posit Science. Often called "the father of brain plasticity," he is one of the scientists responsible for our current understanding of brain change across the lifespan. For nearly five decades, he and his colleagues have conducted seminal research defining the functional organization and rehabilitation of the brain. Dr. Merzenich has published more than 150 articles in leading peer-reviewed journals and has received numerous awards and prizes for his work. He and his work have been highlighted in hundreds of books, news articles, films, and television programs. He earned his bachelor's degree at the University of Portland and his PhD at Johns Hopkins.

To learn more, visit www.soft-wired.com.